Sustainable Energy Education in the Arctic

This book examines the nature of the 'energy curriculum' in Arctic Higher Education and provides invaluable data and new models to assess levels of Sustainable Development Literacy.

Drawing on course mapping conducted in Higher Education institutions across the Arctic, Arruda looks at the nature, structure, and design of the Arctic Higher Education curriculum in order to assess levels of Sustainable Development Literacy and considers the extent to which Arctic Higher Education courses align to UNESCO Education for Sustainable Development (ESD). Using data from four key case studies in Norway, Canada, and the US, and applying a framework drawn from different knowledge systems (Traditional Knowledge and Western educational system), she analyses the different educational approaches and pedagogies used and specifically considers how Higher Education in this region can contribute to the accomplishment of Sustainable Development and the Sustainable Development Goals. The book concludes by proposing new models to assess Higher Education adherence to ESD and outlines how a culturally inclusive curriculum can invite different groups of people to engage in a meaningful Sustainable Development debate, learning experience, and knowledge application.

This innovative volume will be of great interest to multicultural students, scholars, and educators of Sustainable Development, climate change, energy, Arctic studies, and global Higher Education across the Arctic and non-Arctic nations.

Gisele M. Arruda researches, lectures, and supervises at Coventry University London, UK, and is the editor-in-chief for *Arctic, Energy and Climate Change International Journal* by Anvivo.org. She is a principal researcher in circumpolar studies in the fields of energy, Arctic, climate change, environment, and society and has been an advisor/expert for governmental and non-governmental organizations as a strategist by influencing policymakers/stakeholders and supporting climate negotiations (UNFCCC COPs). She is involved in international publications, conferences, and grants. She is the editor of *Renewable Energy for the Arctic: New Perspectives* (Routledge 2018).

Routledge Explorations in Energy Studies

Energy Poverty and Vulnerability
A Global Perspective
Edited by Neil Simcock, Harriet Thomson, Saska Petrova and Stefan Bouzarovski

The Politics of Energy Security
Critical Security Studies, New Materialism and Governmentality
Johannes Kester

Renewable Energy for the Arctic
New Perspectives
Edited by Gisele M. Arruda

Decarbonising Electricity Made Simple
Andrew F Crossland

Wind and Solar Energy Transition in China
Marius Korsnes

Sustainable Energy Education in the Arctic
The Role of Higher Education
Gisele M. Arruda

Sustainable Energy Education in the Arctic
The Role of Higher Education

Gisele M. Arruda

First published 2020
by Routledge
2 Park Square, Milton Park, Abingdon, Oxon OX14 4RN

and by Routledge
52 Vanderbilt Avenue, New York, NY 10017

Routledge is an imprint of the Taylor & Francis Group, an informa business

© 2020 Gisele M. Arruda

The right of Gisele M. Arruda to be identified as author of this work has been asserted by her in accordance with sections 77 and 78 of the Copyright, Designs and Patents Act 1988.

All rights reserved. No part of this book may be reprinted or reproduced or utilised in any form or by any electronic, mechanical, or other means, now known or hereafter invented, including photocopying and recording, or in any information storage or retrieval system, without permission in writing from the publishers.

Trademark notice: Product or corporate names may be trademarks or registered trademarks, and are used only for identification and explanation without intent to infringe.

British Library Cataloguing-in-Publication Data
A catalogue record for this book is available from the British Library

Library of Congress Cataloging-in-Publication Data
Names: Arruda, Gisele M., author.
Title: Sustainable energy education in the Arctic : the role of higher education / Gisele M. Arruda.
Description: Abingdon, Oxon ; New York, NY : Routledge, 2020. | Series: Research and teaching in environmental studies | Includes bibliographical references and index.
Identifiers: LCCN 2019031609 (print) | LCCN 2019031610 (ebook) | ISBN 9780367376734 (hardback) | ISBN 9780429355547 (ebook)
Subjects: LCSH: Education, Higher–Arctic Regions. | Sustainable development–Study and teaching (Higher) | Sustainable Development Goals.
Classification: LCC LA2277.5 .A77 2020 (print) | LCC LA2277.5 (ebook) | DDC 378.98–dc23
LC record available at https://lccn.loc.gov/2019031609
LC ebook record available at https://lccn.loc.gov/2019031610

ISBN: 978-0-367-37673-4 (hbk)
ISBN: 978-0-429-35554-7 (ebk)

Typeset in Goudy
by Integra Software Services Pvt. Ltd.

Contents

List of figures vi
List of tables viii
Acknowledgements x
List of abbreviations xi
Foreword xiii

1 Introduction: energy as the genesis of all human development and learning 1

2 The Arctic context: energy, development, education 28

3 The influential theories for the Arctic Higher Education – Education for Sustainable Development (ESD) 79

4 The nature of the Arctic energy curriculum 108

5 The structure of the Arctic energy curriculum in the case studies 137

6 Curriculum design and review in the four case study institutions 153

7 Levels of energy literacy and adherence to ESD: Arctic Citizenship Model 162

8 Indigenous and non-indigenous students' perspectives 184

9 The relevance of the UN SDGs to the Arctic HE and curricula 210

10 Multidisciplinary and multicultural perspectives of a sustainable energy system in Arctic Higher Education curricula: analysis, discussion, and conclusion 218

Index 256

Figures

1.1	Boundaries and definitions for the Arctic and Subarctic regions according to the Programme for the Conservation of Arctic Flora and Fauna (CAFF)	12
1.2	Internal triangulation	22
2.1	Estimated undiscovered Oil and Gas Arctic basins	30
2.2	Location of Tuktoyaktuk and Ulukhaktok in the Inuvialuit Settlement Region, NWT, Canada	40
2.3	Sami groups in Norway	45
2.4	Indigenous Peoples and Languages of Alaska	47
2.5	Seven drivers of the Triple Bottom Line responsible for the seven sustainability revolutions	55
2.6	Five pillars of ESD	64
3.1	UNESCO Global Citizen Education Framework. Core dimensions of Global Citizenship	96
4.1	General (top) and Detailed (bottom) Indigenous Education Pattern – Course Objectives of Glacial University (GU)	126
4.2	General (top) and Detailed (bottom) Pattern – Course Objectives of Borealis University (BU)	129
4.3	General (top) and Detailed (bottom) Pattern – Course Objectives of Juniper University (JU)	131
4.4	General (top) and Detailed (bottom) Pattern – Course Objectives of Ice University (IU)	132
4.5	Key objectives of the courses as per lecturers' interviews and coding process. The codes that were agreed across all participants (1–6 from Table 4.2) appear in the centre, but this figure is to show the diversity of other objectives mentioned by the faculty members (lectures)	133
4.6	Different modes of learning varying from theoretical to practical	135
7.1	Analytical result based on the Innovative Arctic Pedagogical Model of Energy Literacy applied to case studies. Comparative assessment produced based on the presence or absence (frequency) of the course components from subscales 1 to 14 shown in Table 7.2	170

7.2	Results from the application of the 'Five Types of ESD' analytical model to case studies based on the presence or absence of the components in courses/curricula	182
8.1	The academic characteristic of the courses according to interviewed students scaled by percentage, where 50% means 50% of the course is considered to be vocational and 50% academic; and 95% means 95% of the course is academic with only 5% vocational. The names in black represent the non-indigenous students while the names in grey represent the indigenous students	202
9.1	How well do you know the Sustainable Development Goals?	214
9.2	KNOW: Knowledge of Global Goals and Sustainable Development Goals	215
10.1	Comparative overview on charts summarizing the main course objectives in the four case studies	233

Tables

1.1	Criteria applied to score the most pertinent areas selected for the study	14
1.2	Dimensions of 'systemic thinking' (Senge 1990, p. 68) of seeing, knowing, doing adapted from Sterling (2003); Sterling et al. (2005)	16
2.1	Northwest Territories First Nations and Yukon First Nations	46
2.2	The policy dimension and background of ESD	58
2.3	Two sides of ESD	61
2.4	Three types of ESD	62
3.1	Core capabilities according to Nussbaum (2011)	89
3.2	Pedagogical model designed based on DeWaters and Powers (2011, p. 1700)	93
3.3	DeWaters and Powers (2011) pedagogical model adapted by foundations from St. Clair (2003), Hein (1991), and Mezirow (2009)	94
3.4	Global Citizenship Education: topics and learning objectives	96
3.5	Two sides of ESD	101
3.6	ESD1 and ESD2 – positives and negatives	101
3.7	Three types of ESD based on Scott and Gough (2003, pp. 113, 116)	102
4.1	A summary of the lecturers' code names, educational institutions, countries and education environment	109
4.2	The most influencing factors of course objectives in the case study universities. The objectives are sequenced so that those mentioned by the largest proportion of participants appear first	110
4.3	Features of Education in the case studies shown as a continuum from theoretical to applied knowledge	134
5.1	Curriculum characteristics based on models of the research-teaching nexus by Healey (2005)	138
7.1	Arctic Pedagogical Model of Energy Literacy based on cognitive, affective, behavioural and capability components, and findings. The italics represent the findings from this study that are new, whereas citations indicate those drawn from the existing literature	163

7.2	Innovative Arctic Pedagogical Model of Energy Literacy showing the relevant educational components present in the courses under analysis based on curricular objectives	165
7.3	ESD1 and ESD2 positive and negative aspects based on Vare and Scott (2007, pp. 3, 4), including findings in italics	172
7.4	ESD adherence model based on Scott and Gough (2003, pp. 113, 116). Three types of ESD and notes related to Arctic data collection in the four case study universities and curricula. The words in italics and bold represent the elements from the findings acquired during the Arctic research	173
7.5	Arctic Citizenship Model. Five types of ESD adapted framework from Scott and Gough (2003, pp. 113–116) and Haigh (2014) by adding up the findings from Arctic data collection	175
7.6	ESD Adherence Parameters Analytical Model categorizing five types of ESD applied to the four case studies	178
8.1	A summary of the students' code names, educational institutions, countries, and institution population type	184
10.1	Features of education in the case study universities	234

Acknowledgements

Thanks to all the Arctic people I have met along this journey for the learning and inspiration to my soul.

All of them are major contributors to our sustainable future of knowledge and prosperity.

List of abbreviations

AACHC	Alaska Arctic Council Host Committee
AHDR	The Arctic Human Development Report
ASM	Arctic System Model
BA	Bacharelate
CAVIAR	Community Adaptation and Vulnerability in Arctic Regions
CCHRC	Cold Climate Housing Research Center
CO2	Carbon Dioxide
CSR	Corporate Social Responsibility
EALAT	Reindeer Herders Vulnerability Network Study
EIA	Energy Information Administration
EIA/DOE	Energy Information Administration/US Department of Energy
EL	Energy Literacy
ELOKA	Exchange for Local Observations and Knowledge of the Arctic
ESD	Education for Sustainable Development
FAO	Food and Agriculture Organization of the United Nations
FNIGC	First Nations Information Governance Centre
GDP	Gross Domestic Product
GHG	Greenhouse Gas Emissions
GIS	Geographic Information System
HE	Higher Education
HEFCE	Higher Education Funding Council for England
ILO	International Labour Organization
LNG	Liquefied Natural Gas
MDGs	Millennium Development Goals
NETL	National Energy Technology Laboratory
OCS	Outer Continental Shelf
OECD	Organisation for Economic Co-operation and Development
SATVA	Student and Teacher Value Assimilation
SD	Sustainable Development
SDGs	Sustainable Development Goals
SDL	Sustainable Development Literacy
SDWG	Arctic Council Sustainable Development Working Group
SEL	Sustainable Energy Literacy

SES	Social Ecological Systems
TBL	Triple Bottom Line
TEK	Traditional Environmental Knowledge
TK	Traditional Knowledge
UNDRIP	The United Nations Declarations on Indigenous Peoples
UNESCO	United Nations Educational, Scientific and Cultural Organization
UNFCCC	United Nations Framework Convention on Climate Change
WRH	Association of World Reindeer Herders

Foreword

This comparative study aimed at examining whether the multidisciplinary and multicultural perspectives of a sustainable energy system are being addressed by Arctic HE curricula in times of carbon-constrained activities. The original contributions to knowledge of this book are the Arctic Pedagogical Model of Energy Literacy, ESD Adherence Parameters, and the Education for Arctic Citizenship analytical models, created to assess levels of adherence to SD and energy literacy. The integration of different theoretical approaches like Complexity Theory, Capability Theory, Global Citizenship Education, and Constructivism provide the theoretical foundations of this research contributing to enhance understanding about the nature of the 'energy curriculum' for Arctic Higher Education by providing data on the levels of energy literacy and of adherence to Education for Sustainable Development (ESD). This book uncovers these educational elements and shows that the energy transition period will require well-equipped citizens engaged in a sustainable energy vision. The research outcomes revealed that a traditional Western education, emphasizing the economic pillar of the TBL, does not provide the necessary adaptability mechanisms to face the scale of the current Arctic and global transition. A culturally inclusive curriculum that includes contextual, multicultural, and multidisciplinary components with higher levels of adherence to ESD, energy literacy and, geocapabilities seems more adequate to invite different groups of people to engage into a motivating and meaningful dialogue, learning experience, and knowledge application that a sense of regional and global citizenship will require.

1 Introduction

Energy as the genesis of all human development and learning

1.1 Energy as the genesis of all human development and learning

This chapter introduces this book by explaining the evolution of humans' relationship with energy as a means to introduce sustainable energy systems and the key concept of Sustainable Development (SD).

The starting point of this research and educational reflection on the Higher Education curriculum involves energy and the importance it has to human development since the first stages of human existence. The word first appeared in English in the 16th century based on a Greek word used by Aristotle, but the concept of energy, in the modern sense, appeared in the early 1800s developed by scientists to try to describe their observations about the behaviour of phenomena related to the transfer of heat, the motion of bodies, the operation of machinery, and the flow of electricity. The standard scientific definition is relatively recent dating from the 19th century and relates to the capacity to do work, to move an object against a resisting force (Everett et al. 2012).

Energy has strong links to natural sciences as a physical process and it is linked to the most vital biological processes in the food chain, if we consider that algae, plants, and cyanobacteria capture light from the Sun to metabolize energy. After being captured as solar energy, it flows through the food chain assuring organisms' survival. In other words, all forms of life depend on energy to live and to develop their vital functions, inclusive of human beings as part of the Earth's ecosystem.

The history of energy is the history of human development since the discovery of fire 400,000 years ago when primitive human societies developed an understanding of energy by using fire to heat, to cook, and as a fuel. Other uses of energy involved the production of coal and metals despite the general per capita energy use remaining at a constant level until the Industrial Revolution. This process can be seen in parallel with the agricultural revolutions. In the first agricultural revolution, shifting cultivation was a common, as well as low-energy method of farming where 'slash-and-burn agriculture' used to be practised. It basically consisted in controlled use of fire in pre-determined places where the original vegetation was burnt and a layer of ash from the fire was left on the ground to promote soil's fertility. The so-called subsistence farming opened space to a different style of cropping, more intensive, more communal, and more organized. This model of farming had a significant impact on turning human society stratified with important impacts on wealth distribution (Waters 2007, p. 48).

It was the second agricultural revolution together with the Industrial Revolution that represented the huge human transition for what we know today. The agriculture considered beyond subsistence, mechanized, commercial, and able to generate the necessary surplus to feed people in urban centres and wars gave stage to the technical innovation capable of creating new environments, new demands, new demographic scenarios, and new models of economy. Energy is an element in constant transition that causes the human being to be in constant transition either. This transition requires better understanding, study, and mapping (Arruda 2018a, p. 115).

In this movement of human progress and creation of an industrial economy, the third agricultural revolution or 'green revolution' emerged as an international attempt to combat hunger by enhancing crop performance through the combustion engine, mechanization, biotechnology, and modern financial practices to enable poor countries of the world to access the necessary capital and technology to attend the need of a growing population. This progressive development was the great mark of an intensive use of energy, not only to move agricultural machinery but to move the whole world economic system created from these revolutions and involving transport, heating, cooling, production, industrial systems, and supply chains (FAO 2017, p. 4).

This is the great historical mark for our knowledge journey when studying energy transition and human development; the time when the steam engine was created and made it possible to convert chemical energy into motion energy; and when the concept of energy efficiency emerged and the fossil fuels started to move an industrialized society. Immediately after the invention of the 'horseless carriage' (automobile) and the lamp bulb, the world could understand and dominate energy by the generation of combustion process and electricity, triggering large-scale energy production.

After the discovery of fire, this historical mark represents the beginning of a human intensive energy production and use that revolutionized human society's life, economics, well-being, environmental management, as well as the future of earth's ecosystems. Coal remained the main used source of energy until the middle of the 20th century when it was overtaken by petroleum. Meaning 'rock oil', it is the most efficient source of energy discovered so far. The essential pre-conditions for petroleum formation include the abundance and diversity of plant and animal life and a seabed environment of dead organic material breakdown by bacterial oxidation. The burial of such hydrocarbon-containing plant and animal material under sea-borne mud and silt 'preserves' the material for the next stage of the petroleum-forming process that consists in compression, high pressure and temperature provoking a chemical maturation and formation of petroleum stored in geological traps (Everett et al. 2012, p. 214). The start of a modern commercial extraction dates from 1859 when Edwin Drake started drilling in Pennsylvania (Maugeri 2010). Along the last 200 years, crude oil could produce a range of by-products from natural gas to waxes and bitumen through highly technological industrial processes of distillation and refining, and its uses revolutionized other sectors like transportation, oil-fired turbines, the natural gas industry, and the alternative fuel processes.

This energy source discovery was not only important for its practical applications and the energy revolution it caused, but it was also extremely important for triggering the evolution of modern industrial landscapes with direct effects on human societies, economic development, and global human geography (see pp. 7, 8, 19). Human societies were impacted by the geographical variations and changes in the economic activities marked by the evolution in the use of natural resources through historic levels of economic activity. This historic human trajectory was marked by the migration of primary economic activities (farming, forestry) to secondary economic activities related to conversion into more valued products (manufacturing) to tertiary economic activities (trade, financial services), to quaternary economic activities concerning the generation, processing, and transmission of information, intellectual services and business, management, and entrepreneurship (Dicken 2003). This shift has significant implications to understand the status of the Sustainable Development process.

The importance of energy discoveries along this societal evolutionary process is undeniable, a global resource system feeding economic activities, triggering technological advances, and new political scenarios. Energy is the essence of the global capitalist manufacturing system and it dictates the international patterns of production and consumption, disparities between demand and supply, price fluctuations, and the different levels of development among geographical areas of the world. Energy moves the globalizing economy and the globalized world by dictating new patterns of economic growth and human development in several geographical areas of the world depending on the abundance or scarcity of energy resources used in different energy systems (OECD 2011, p. 9).

In this sense, it is crucial to understand the concept of energy systems. A system is a set of interconnected entities that are part of an integrated whole, and the key element of 'systems thinking' is to consider that it has a system boundary for analytical purposes. The boundary between the system and the environment becomes crucial to define rules to safely operate the system. This is the main concern when we discuss the needs of expanding production and use of energy in relation to specific geographical energy systems.

Energy expansion or demand brings a certain tone of concern, as the characteristics of today's energy system are the growth in world primary energy consumption since 1850, parallel rises in world population, world Gross Domestic Product (GDP) (Everett et al. 2012, p. 3), and global levels of Greenhouse Gas Emissions (GHG). This energy expansion trend influences the current global energy system consisting of production and use of three main sources of energy, i.e. fossil fuels, nuclear energy, and, in lower percentage, renewable sources. Fossil fuels, or hydrocarbons, due to their composition of carbon and hydrogen, have been a very attractive energy source mainly related to flexibility in transformation, efficiency, and high combustibility. Coal, oil, and gas have been used for more than two centuries (Hilyard 2012) and continue to be the base of the current global energy system despite the clear signals of depletion and decline in the main supply basins producing since the beginning of the 20th century.

Another relevant source of our current energy system is nuclear energy derived from fission of the nuclei of heavy isotopes such as uranium-235 and plutonium-239 (OECD 2001). Despite being, currently, a controversial energy source with a vast history of case studies related to nuclear accidents and environmental impacts, nuclear power plants continue to be built and reactivated as a result of the dramatic rise in the energy demand with specific projections for 2030 and 2050 by the main energy agencies from developed countries.

The above changing scenario consisting in the rise of energy demand associated to the growth in world population and the world GDP per capita broadly justifies the necessity of energy expansion systems via renewable resources that can regenerate and respect the system boundaries by promoting environmental resilience (Arruda 2018b, p. 112).

In order to understand the concept of a 'sustainable energy system', it is essential to understand what Sustainable Development and sustainability are. Both were first presented by the United Nations' Brundtland Commission report entitled 'Our Common Future' from 1987, when sustainability was identified as the practical function of 'Sustainable Development' defined as 'Development that meets the needs of the present generation without compromising the ability of future generations to meet their needs' (UNWCED 1987, p. 43). In terms of energy systems, being sustainable means not to deplete or decline with constant use; not to emit hazardous levels of gases and pollutants to affect human and ecosystems' health; and not to perpetuate social injustices (Everett et al. 2012, p. 10). These are important principles that require unfolding and motivated the present study through their importance to Higher Education in terms of Sustainable Development Literacy (SDL). These concepts and principles are better understood when a systematic approach in relation to energy studies is adopted by the implementation of energy literacy in the curriculum as a fundamental component of SDL. For the Arctic it has an even higher significance due to the sense of urgency to tackle climate change and the relevance of understanding how to structure and operate a sustainable energy system. Energy literacy is essential to realize the sustainable energy vision by reorienting patterns of production and consumption.

The Brundtland Report (UNWCED 1987) also introduced two new components in the energy discussion as a component of 'time' when referring to future generations and a component of 'transition' or adaptation not only in relation to energy systems but also in relation to general human systems of existence. The time component (time dimension) is important to understand the long-term approach to energy production and use, while the transition component is essential to assure that the long-term approach is achievable through non-exhaustible sources of energy. These two components point out to the importance of the renewable energy and its modalities for shaping new futures (Arruda 2018b, p. 115). For centuries, human societies from different parts of the world have used energy from water wheels (hydropower) and windmills (wind energy), but only recently hydroelectricity, wind energy, and solar energy started to be considered in the energy matrix as important sources for attending the high global energy demand (Boyle 2012).

In the last 30 years, due to the recurrent discussions about energy efficiency programmes, climate variations, and the risks associated to the rise in the global energy demand with serious projections to the future, a different approach in relation to energy production and use started to be debated within the industry and academically (MacKay 2008; REN21 2017). The intensification of the debate regarding renewable energy resources being the base for a sustainable energy system (SES) has triggered important technological developments contributing to make these alternative energy options more accessible, despite the costs of producing renewable resources continuing to put pressure on regional and local sustainability policies. However, this trend seems to be in rapid change due to the recent scientific evidence presented in relation to the high concentration of Greenhouse Gases (IPCC 2000) exactly dating from the industrial revolution era. Apart from that aspect, the huge projected energy demand, associated to the demographic growth and GDP per capita, turned to be a key imperative to accelerate the energy transition and the so-called 'decarbonization of the economy' (Stern 2007).

The complexity of the climate system in the Arctic inspired modelling and prediction studies that consider the interactions among other systems like atmosphere, oceans, cryosphere, land surfaces, and biosphere (Lemke & Jacobi 2012, p. 350) over spatial and temporal scales. Since 1994 when the Arctic Climate System Study (ACSYS) was launched, representation in models of interactions among these systems and patterns of radiation and clouds continue to dominate Arctic climate change research (Dethloff et al. 2012, p. 326), generating simulations of important regional physical processes and patterns in light of palaeoclimatic records (Alley 2003, p. 1831).

These patterns observed through the glaciers velocity, the accumulation zone of the Greenland Ice Sheet, and the fluctuation on the surface mass balance originated live and dynamic natural systems representing an important source of data and knowledge to be interpreted by the international academic community, but it is also an important part of the local academic investigation in Higher Education in order to orient the local response to this important driver of change that also presents important impacts on other local systems.

This is a highly complex process and subject that requires in-depth interdisciplinary knowledge, competences, and practical learning as discussed in more detail throughout this book.

1.2 Background – energy in the geographical landscape of the Arctic

Based on the historic perspective of energy development, the major discoveries of different fuels, and the developments of energy conversion technologies, significant geographical impacts on human development take place due to energy production and use. It is true that energy can be studied by different disciplines like physics, engineering, business, and management; however, it is irrefutable that energy is a geographical concept as its production, use, and access vary according to the geographical region and it directly implies local geographical levels of economic and human development (Bridge et al. 2013, p. 331).

This research focuses on energy as a component of human geographical studies, aiming at assessing the implications of energy production and use on a highly sensitive area that is the Arctic. It also focuses on how Higher Education is promoting learning experiences related to energy and Sustainable Development across the Arctic region.

The ways in which societies secure energy and transform it to do useful work exert a powerful influence on their economic prosperity, geographical structure, and international relations (Bridge et al. 2013, p. 331). This assertion brings on board a number of elements to consider as by examining the energy transition as a geographical and learning process, and by assessing the myriad of impacts of energy production and use in an extremely sensitive area like the Arctic, it can elucidate several relevant implications to the non-Arctic world. The same parallel can be made in relation to the educational side of energy.

Environmental impacts of energy production and use do not respect frontiers and affect equally Arctic and non-Arctic geographical landscapes. At this point in human development, 'ensuring the availability and accessibility of energy services in a carbon-constrained world' will require 'developing new ways, new geographies', new education of 'producing, living and working with energy' (Bridge et al. 2013, p. 331). According to Bridge et al. (2013) in his 'Geographies of energy transition: Space, place, and low-carbon economy', the geographical implications of this 'new energy paradigm' are not well defined, and a range of quite different geographical futures are currently possible (Bridge et al. 2013). This seems an extraordinarily important reason to study energy literacy in the Higher Education curriculum.

Energy is a highly interdisciplinary subject, bringing on board different facets like economy, geopolitics, bio-geo-physical effects, and socio-environmental impacts concomitantly. Observations from my own practice of education reveal that, in general, students do not have the deep understanding of the relevant concepts (energy efficiency, renewable energy, Sustainable Development, energy literacy) and the necessary interdisciplinary perspectives to operate the energy transition and to make well-informed decisions in practical situations. Concepts like Sustainable Development and its relationship to energy are not sufficiently disseminated in undergraduate courses in Higher Education curricula (Cotton et al. 2009). Theory and practice seem not to be conveniently connected to promote the interdisciplinary understanding the topic requires.

Considering the complexities of the current global development models, based on fossil energy production and use, this book argues that a sustainable energy future is highly influenced by a deep understanding of current energy systems, sustainable energy system, and Sustainable Development. An energy future depends on an innovative and meaningful energy education through a holistic Higher Education curriculum. However, the energy transition inspiring a human developmental transition also requires an education transition on global scale not only in the Arctic. If this approach starts in the Arctic, it may prevent or minimize the severe impacts of energy production and use to this specific region's ecosystems, by triggering several other environmental changes and dynamics that can affect the whole world. The present and future generations mentioned in the Brundtland report definition of Sustainable Development will need to have the necessary knowledge and adaptive skills to understand the current scenario and operate the changes of paradigm involving mainly the management of negative environmental impacts and the integrative management of new energy technologies in an efficient way considering a multitude of old and new variables. The reconfiguration process involving current spatial patterns of economic and social activity will require a systematic educational approach in relation to environmental, human, and energy systems at different levels, considering undergraduate and postgraduate courses and curricula. This book uncovers these educational elements and shows that the energy transition period will require well-equipped citizens and professionals aware of the interdisciplinary and multicultural challenges of energy security and efficiency, climate change management, socio-environmental impacts, and Sustainable Development (SD).

Adaptation seems to be the key word for the future. It is a powerful word based on a challenging and impermanent reality of continuous complex interactions between the natural and social structures on Earth. These interactions are better understood when we study them through a geographical perspective (Arruda & Krutkowski 2017), as some of these interactions, depending on the geographical area considered, can present dramatic consequences or create extraordinary geo-capabilities for learners. Considering that geography involves the holistic study of the interactions among human and natural systems (Norton & Mercier 2016) and the contemporary world presents a series of changes that will require severe physical and social adaptation, it is possible to argue that, the geographer's work is more important than ever. Geography is uniquely positioned to link different disciplines together and it provides the basic building blocks for the study of resource use and Sustainable Development, within the human environment. As resources are limited and finite, humans as agents of change must employ strategies that allow an efficient and durable use of the available resources (Aikins 2014, p. 261; Huckle 2002, p. 65). The spatio-temporal dimensions of sustainability call for geographical approaches to be able to understand the dynamics, complexity and interactions in various scales (Firth 2011, p. 14; Grindsted 2013, p. 18). The geophysical aspects of our world are naturally interlinked. With globalization, the human aspects of our existence are even more integrated resulting in a range of collective opportunities, risks, and responsibilities as a civilization, as citizens.

These integrated geo-bio-physical and social systems of existence require an accurate perception on the part of the human beings, as the interconnections of their causes and consequences require knowledge and informed-decisions. Adaptation is simply part of this learning dynamic, requiring even more sophisticated levels of perception, understanding, skills, and knowledge in order to anticipate negative impacts on the geographical human landscape.

More accurate perception, knowledge, and skills can be substantially developed by recognizing the important role of education in our contemporary world demanding constant and more accelerated rates of adaptation. It is undeniable that a geographical perspective will permeate our present and will shape our future as well as our capability of understanding and acting in face of the world's dynamic complexity.

The basic premise of energy literacy as a key component of 'Sustainable Development Literacy' is that human and natural systems are dynamically interdependent and cannot be considered in isolation when addressing critical issues such as climate change and resource depletion. The element of change considered in this research is the understanding of what energy production/use entails in the environmentally fragile Arctic, and the capacity to communicate this learning through a geographical viewpoint at Higher Education (HE) level. Observations from the last 18 years of teaching practice reveal that HE students seem not to have a clear understanding of SD and energy. This book therefore aims to explore this in the context of Arctic Higher Education, the aims and objectives are outlined next.

1.3 Focus, objectives, and contributions

This book argues that our modern society is facing an energy crisis owing to an escalation in global energy demand, the insecurity of supply sources, continued dependence on fossil fuels for energy generation and transportation, and an increase in world population. These are factors responsible for depleting natural resources and a steady increase of carbon dioxide emissions, which experts like Helm (2007) and Stern (2007) believe is responsible for increasing average global temperatures and triggering cyclical climatic variations.

The Arctic, the ultimate case study (Arruda 2015), is the last frontier for energy exploitation and home of an unexplored cluster of natural and human resources. It is known that 20% of the world's energy resources in terms of fossil fuels like oil, gas, and coal are in the Arctic (USGS 2008). The U.S. Geological Survey estimates that 23 billion barrels of oil and 108 trillion cubic feet of natural gas offshore are recoverable from the Outer Continental Shelf (OCS) (USGS 2008). The extraction of these resources represents an opportunity and a risk – a change that will require socio-environmental management and a new paradigm of governance.

Contemporary and rapid social change in the Arctic region also ignites new ideas and practices of modernity contrasting locally situated practices (Traditional Knowledge – TK), accelerating the fragmentation and dispersal of modernity into constantly proliferating 'multiple modernities' triggering 'distorted' or 'divergent' patterns of development, and re-assembling what is often designated as 'tradition' (Arce & Long 2000, p. 1).

Climate change has a significant role in the Arctic socio-environmental changes as the melting ice triggers new debates over territorial sovereignty of Arctic coastal states (Berkman & Young 2009). Indigenous communities in the resource-rich areas of the Arctic are increasingly exposed to severe climate change impacts as well as the external pressures of development advocated by governments and their industry partners (Arruda 2018a). With the discovery of vast energy resources in previously inaccessible areas of the Arctic, the governments of littoral states are taking new measures to assert their territorial sovereignty over frozen land and the newly opened waterways turn into trade routes (Arruda 2018a; Elferink 2011). Climate change and modernization have thus become two intrinsically linked forces that severely alter the context in which the indigenous populations sustain a livelihood (van Voorst 2009).

Teaching and research experience in Energy Governance and Environmental Management for Sustainable Development provided the invaluable opportunity to identify international undergraduates' limited intercultural and multidisciplinary understanding about SD, SDGs, energy and climate change adaptation, and consequently, a low curriculum adherence to ESD. Interdisciplinary participatory pedagogies, 'real world' research, and a systemic view on sustainability provide a way to contribute to the transformative aim of HE according to Tilbury (2011) and Haigh (2014).

The focus of this comparative study is to investigate whether the multidisciplinary and multicultural perspectives of a sustainable energy model are being addressed by Higher Education curricula in the Arctic in times of carbon-constrained activities. This study also contemplates how HE can enhance mechanisms to integrate 'energy' and 'climate adaptation' into the undergraduate experience by adopting different perspectives and innovative pedagogies to engage students in this sustainable energy vision. In this sense, it is crucial to define the structure of the energy curriculum from four case study universities in the Arctic; to analyze the processes through which these curricula are 'made' and reviewed; and to define the structure of the energy courses offered in the selected universities and compare them according to their adherence to ESD.

This research was originally conceived through expeditions made to Arctic locations (Norway, Canada, Alaska, Iceland, Greenland, and Faroe Islands) and direct contact with local communities and local energy production areas. The experience has shown that it would be beneficial to improve societal understanding of energy production and use and Sustainable Development in highly sensitive areas of the Arctic.

Sustainable Development and education have an intrinsic, interdependent relationship with each other, largely because no sustainable society can operate without high-quality education as well as human development being intrinsically linked to education standards (AHDR 2014; Poppel 2015, p. 67). Curricula which only focus on key competencies for the world of paid employment are deficient (Marsh 2009, p. 7) because learners are persons not only industry professionals. The developmental model of curriculum planning suggests that we should look to a particular view of humanity and of human development, including social development and cultural components (Kelly 2009, p. 88) because cultural difference contributes to the managerialist and organizational goals of development (Radcliffe 2006, p. 237).

This study enhances our understanding about the nature of the 'energy curriculum' for Higher Education in universities within the Arctic Region by providing data on the levels of energy and Sustainable Development literacy by examining the characteristics of the current energy education curricula in the Arctic. It provides clarification on the processes through which these curricula are founded, 'made' (designed), and reviewed. Ensuring energy security in a carbon-constrained world involves a broader understanding of the relationships between energy, climate change, and development to inform decision-making for the Higher Education sector.

This comparative study investigates whether the complex and intercultural perspectives of a sustainable energy literacy are being addressed by Higher Education curricula in the Arctic. It informs human development theory (Nussbaum & Sen 1993) by looking at the capabilities which the curriculum is designed to deliver and triangulates accounts from programme documents, with those of academics' and students' interviews.

The focus and commitment of this study is to present the participants' perspectives on sustainable energy literacy for the Arctic, which are deeply dependent on their views of Sustainable Development in a sensitive geographical region. Despite application in other disciplines, energy is originally a geographical concept, the main premise of this study (Bridge et al. 2013, p. 331) which leads us to explore the interconnections of energy, geography, and education according to a qualitative paradigm.

In this study, energy is seen from different perspectives related to the interfaces with societal structures. Energy, as a fundamental interface, triggers interest in the processes shaping energy education through the energy curriculum. The qualitative research methods used here intend to elucidate individual and collective experiences and social processes in a particular human environment within a variety of conceptual frameworks (Hay 2016, p. 5).

Societal structures are worth analyzing because structures represent frames that can be used to constrain, or silence, individuals and groups. In this sense, the qualitative research methods based on participants' perspectives provides, what Johnson et al. (2007, p. 118) calls, the 'anti-colonial geographies' in action. In this way, the research may reveal unheard voices, unknown perspectives within largely consolidated concepts that are not being realized in practice.

The approach has been structured to give voice to participants, in order to better understand multicultural and contextual characteristics related to the areas elected for the study, the educational institutions, the targeted departments, and the sample and curricular approaches in high, low Arctic, and subarctic areas.[1]

The field work in different latitudes emphasized the point made by Evans et al. (2014, p. 179) when arguing that indigenous approaches to research are fundamentally rooted in the traditions and knowledge systems of indigenous peoples themselves. Indigenous viewpoints influence methodology, pedagogies, and curriculum in ways that contrast to the mainstream methods and curriculum used in Western-style educational institutions. This point will be widely examined in Chapter 4 of this study.

The study considered the importance of developing a good relationship with different audiences and the researcher paid special attention to the Indigenous Knowledge presented in the perspectives, stories, experiences, and history of different indigenous peoples' viewpoints. These elements were registered by using their own words according to a constructivist methodology.

1.4 Case study selection

Aimed at better understanding problems and orienting to concrete solutions, case studies are considered a research approach which helps to provide detailed analysis of theoretical concepts or, in other words, which broadens academic understanding about a phenomenon from a real-life

12 Introduction

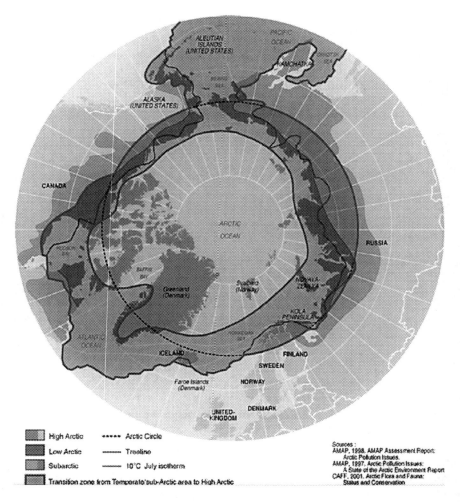

Figure 1.1 Boundaries and definitions for the Arctic and Subarctic regions according to the Programme for the Conservation of Arctic Flora and Fauna (CAFF). Source: (Holberg et al., 2003)

context. Baxter (2016, p. 130) argues that the case study research involves the study of a single instance or small number of instances of a phenomenon in order to explore in-depth nuances of the phenomenon and the contextual influences on and explanations of that phenomenon. In the start of the methodological reflection about the countries to be studied, several options were considered in relation to Arctic areas where energy and education are necessary to promote SD. After a season of field work and initial pilots of the project in places in the High Arctic, Low Arctic, and Subarctic regions (see Figure 1.1), where settlements were visited and direct contact with communities was made regarding local extractive projects and challenges related to energy production and use, it was clear that the development of this topic would be of relevance across the wider region. However, it was impossible to look at every country, so a sample was chosen which would allow findings to be relevant across the Arctic.

Based on the local knowledge formed and the diversity of characteristics associated to these areas, it was possible to design objective criteria to score the most suitable areas for the study. Considering that all Low and High Arctic and Subarctic areas are sensitive in their ways with their specific ecosystems and social contexts, the selected criteria follow three specific levels.

At the first level, the focus was 'Impacts of change'. Climate-change knowledge base, the presence of natural resources, new areas for resource exploration, and the presence of indigenous communities are the fundamental components to define the study target areas due to the socio-environmental impacts caused by those variables in need of a better comprehension by the local community.

At the second level, the criteria consisted of the 'Nature of the existent resources' at stake, as oil and gas, minerals, renewable energy, offshore and onshore, and the potential for energy production and exportation contributed to the selection of areas where energy production and use are part of the local economic activities.

At the third level, the components focused on 'Research & Development' in an attempt to select the Arctic areas involved in knowledge and technology transfer according to international standards and quality levels. The fourth and last level of the criteria considered 'Education & Training' to evaluate ESD adherence in the studied locations, state-controlled curriculum, and the accreditation system practised to evaluate courses, programmes, and institutions.

Therefore, the criteria designed to establish the sample of countries within which case studies were selected were: the presence of natural resources (oil and minerals); the presence of renewable resource projects; resource exploration projects; the presence of high energy mineral export potential; the presence of indigenous populations in the area; state-controlled curricula; levels of research and development; knowledge transfer; and ESD adherence (see Table 1.1). All factors were equally weighted, with a simple tick for each being present, scoring one point. The countries with the highest scores were selected.

Table 1.1 summarizes the geographical areas that obtained the highest scores under the specific criteria used to elect the most representative case studies. Norway (score 18), Canada (score 16), and US (Alaska – score 15) were selected as case studies for this research, and the second group of countries with the highest scores, Iceland, Greenland, and UK (10) were considered countries that would benefit from the study by the application of the results. It also shows the other countries that scored under 10 that

Table 1.1 Criteria applied to score the most pertinent areas selected for the study

	Iceland	Greenland	Norway	Sweden	Finland	Alaska	Canada	Russia	UK
Climate change knowledge-base	✓	✓	✓	✓	✓	✓	✓	x	✓
New Frontier – new areas for resource exploration	x	✓	✓	x	x	✓	✓	✓	x
Natural resources	✓	✓	✓	x	x	✓	✓	✓	x
Indigenous people	x	✓	✓	✓	✓	✓	✓	✓	x
High-energy mineral export	x	✓	✓	x	x	✓	✓	✓	x
Research & development	✓	x	✓	✓	✓	✓	✓	x	✓
State-controlled curriculum	x	x	x	x	x	x	x	✓	x
Education & Training	✓	✓	✓	✓	✓	✓	✓	x	✓
Arctic Policy & Program	✓	x	✓	x	x	✓	✓	x	✓
Oil + Minerals	x	✓	✓	x	x	✓	✓	x	✓
Renewables exploitation	✓	✓	✓	✓	✓	✓	✓	x	✓
ESD adherence	✓	x	✓	x	x	x	✓	x	✓
Offshore/onshore OG resources	x	✓	✓	✓	✓	✓	✓	✓	✓
Sensitive Arctic environment	✓	✓	✓	✓	✓	✓	✓	x	✓
Knowledge transfer	✓	x	✓	x	x	✓	✓	x	✓
Technology transfer	x	x	✓	x	x	✓	✓	x	✓
Accreditation system (European Standards and Guidelines (ESG))	x	x	✓	✓	✓	x	x	x	✓
International standards and quality level	✓	x	✓	✓	✓	✓	✓	x	✓
Score	10	10	18	9	8	15	16	8	10

were not considered relevant for the study because of geopolitical issues related to political instability, politically controlled energy activity and curriculum, lack of reliable database (Russia), and for not being leading areas for energy production (Finland and Sweden). The high scores denote the presence of the critical components triggering rapid Arctic change with impacts on the environment and societies.

An evaluation of the case study areas was completed based on documents explaining the university department structures, course structures, respective energy curricula, and government roles in curriculum design. These aspects of the first phase of studies were decisive to inform the field work plan, considering that the geographical areas elected for the research are highly relevant for being traditional energy production areas in socio-environmentally sensitive areas. The work developed in the universities and respective educational authorities' departments was facilitated, at this stage, by the collaboration of deans, heads of departments, librarians, and officials as part of an existing network developed along two decades of academic and interdisciplinary collaboration.

1.5 Phase 2 – qualitative data collection methods

The second phase comprised field work with a focus on the evidence regarding contrasts and levels of adherence with ESD and energy literacy. Lecturers and students were interviewed, using semi-structured interviews to collect evidence about levels of adherence with ESD and energy literacy. This method of acquiring data was important to reveal specific sets of knowledge, skills, attitudes, and values that shape a sustainable future scenario in the targeted areas as well as key SD issues involved in teaching and learning. The energy literacy levels provided an understanding of the nature and role of energy in their lives. The interviews also provided data informing the interviewees' ability to apply these understandings to answer questions and solve contemporary problems. The use of audio recordings was employed to identify new elements to improve and realize the 'energy vision' for education in the targeted areas. The questionnaires were shaped by elements from the literature, observations, conferences, and 'brainstorming' with stakeholders. Considering that the geographical areas elected for the study present multicultural characteristics, indigenous participants members of the local academic communities as staff and students were invited to participate as interviewees.

1.5.1 Methodological procedure

This study contrasts different participants' perspectives according to their specific geographical contexts in a common system (Higher Education within the Arctic) analyzing patterns of correlation between dimensions of cognition, knowing, and paradigm as per Table 1.2.

Table 1.2 Dimensions of 'systemic thinking' (Senge 1990, p. 68) of seeing, knowing, doing adapted from Sterling (2003); Sterling et al. (2005)

Dimensions	Seeing domain	Knowing domain	Doing domain
Cognition	Perception	Conception	Practice
Knowing	Epistemology	Ontology	Methodology
Paradigm	Ethos	Eidos	Praxis

Central to the task of educators is to conceive the educational work as a system rather than as a set of isolated activities. 'Systemic thinking' (Senge 1990, p. 68), derived from Complexity Theory of Chapter 3, represents an attempt to overcome the linear, mechanistic, and fragmented way of perceiving, thinking, and learning by adopting a more integrative, inter-relational, and holistic style of challenging the current paradigm.

The qualitative paradigm is also a good way of learning about Arctic social context. Quantitative research is useful and valuable, but, in this context, it presents limitations that would dehumanize the participants' experiences and perspectives on social structures. The context of participants' views affects their behaviour, as they are grounded in their history and temporality (Holloway & Galvin 2016, p. 11) which contributes to the elucidation of their social processes.

Qualitative research presents common and relevant characteristics for the study in question, representing a firm justification of the research paradigm as per Holloway and Galvin (2016, p. 10) as follows:

- The data have primacy.
- Qualitative research is context-bound, and it is context sensitive.
- Immersion in the natural setting of the people whose thoughts and feelings they wish to explore – observation.
- Focus on the 'emic perspective', in other words, the views of the people involved in the research and their perceptions, meanings, and interpretations.
- Thick description: to describe, to analyze, and to interpret.
- The relationship between the researcher and the researched is close and based on a position of equality as human beings.
- Data collection and data analysis generally proceed together, and in some forms of qualitative research, they interact.

These characteristics are substantially related to the phenomenology and the relevance of the researched groups' perspectives fitting perfectly the research aims focused on lecturers' and students' constructs in relation to Education for Sustainable Development (ESD) and levels of energy literacy being practised through the curricula. These aspects were explored in depth, thanks to the good relationship and trust built up with universities, communities, and governmental agencies, creating a very stimulating environment for the researcher and the participants.

Researcher and participant are both human beings subject to human experiences (Wilde 1992, p. 240) and subject to sharing these experiences according to an 'empathetic approach' expressed by the term 'Verstehen' or 'understanding something in its context' (Schaefer 2017). Human beings make sense of their subjective reality and attach meaning to it. Social scientists approach people to explore their world within their life context. This worldview applied to a region like the Arctic makes the understanding of human experiences even more important than focusing only on explanation, prediction, and control because it stimulates learning, systems thinking, and a participative reality (Heron 1996).

The focus and commitment of this study is to present the participants' perspectives on sustainable energy literacy for the Arctic, which are deeply dependent on their views of Sustainable Development (SD) in a sensitive geographical region. Despite application in other disciplines, energy is originally a geographical concept, the main premise of this study (Bridge et al. 2013, p. 331), which leads us to explore the interconnections of energy, geography, and education according to a qualitative paradigm.

A qualitative research study may also yield the development of a new theory (Levy 2014, p. 4). In this research, theory is developed inductively out of the research process. The inductive nature of this study derives from the data collected rather than being formed before the data collection process. The study generates data out of which a theory is built. The theory is grounded in the empirical data from the study. Induction into the disciplinary structures of knowledge is important even if we wish to overturn elements of those structures, because understanding the disciplinary structures is a necessary condition for triggering a curricular reflective process.

Grounded theory is an inductive strategy for theory development by analytical comparison of data with data (Brinkmann 2013, p. 37), data with codes, by developing categories turned into theoretical concepts. It was founded as a practical approach to help researchers understand complex social processes, but it is important to remark that grounded theory is not perfect or unproblematic and some educational researchers as Parkhe (1993) and Thomas (2007) believe that due to its genealogy it is inherently 'messy' due to the lack of well-delineated epistemological boundaries. In this sense, adopting grounded theory would require from the researcher a deep knowledge of the subject under study as well as a special attention to the consistency between the method and a particular view of reality.

After the data collection, the researcher applied coding by examining, comparing, conceptualizing, and categorizing data (Strauss & Corbin 1990, p. 61). This research applied grounded theory which enabled an invaluable

18 Introduction

insight into the analysis of energy curriculum-making in a comparative manner (between Canada, Norway, and Alaska).

1.5.1.1 Interviews

The interviews with staff and students had a focus on levels of adherence to ESD and levels of energy literacy. The methodology involved interviews with four lecturers and six undergraduates using semi-structured interviews to collect evidence on levels of energy literacy and on how curricula are made and reviewed. Data collection provided conditions to make a comparative analysis on what extent these curricula meet academic/vocational aims and promote intercultural understanding of energy production and use. Data collection started in Norway at the Glacial University and then proceeded to the Borealis University, then the data collection continued at the Juniper University and, finally, at the Ice University. Qualitative coding of interviews took place in phase 3.

1.5.1.2 Areas covered by the interviews

As the principles of SD call for a comprehensive change in the way society operates, it was important to analyze to what extent curriculum innovation has the potential to develop the skills needed by the Arctic energy development scenario. It is not only a matter of incorporating the principles of SD and environmental awareness into HE, it also involves training educators in sustainability issues, apart from embedding sustainability into the daily practice. This process ensures that updated curricula take economic, social, environmental, and cultural dimensions of SD into account and ESD becomes an integral part of training of leaders in business, industry, trade unions, non-profit and voluntary organizations, and the public services. Other fundamental themes covered by the interviews explored processes of curriculum-making by the different institutions; the geo-capabilities the curriculum stimulate in learners; levels of energy literacy; and sustainability pedagogy in HE for the Arctic.

This form of interviewing was useful to complement the interpretive approach to reveal how people understand their worlds and share meanings about their lives (Rubin & Rubin 1995, p. 17). The geographical location of the interviews differed according to the institutions researched as the interviews occurred at the premises of the studied educational institutions. The researcher travelled to Norway, Canada, and Alaska to conduct the interviews, as it was important to establish a deep conversation and interpret emotions and impressions from the face-to-face contact in order to interpret multicultural body language.

Social science investigation is highly permeated by language, culture, and dialogues associated to human life. Conversations are a rich and indispensable source of knowledge about individual and collective aspects of specific contexts. According to Taylor (1989, p. 36), we are immersed in 'webs of interlocution' due to the conversational characteristic of the human being. As we are linguistic creatures, language is an essential component of the conversational process of knowing (Munhall 2007). Under a multicultural perspective, this interchange becomes even more relevant because cultures are produced and reproduced in dialogues among their members (Mannheim & Tedlock 1995, p. 2).

When immersing in the 'real' world of the participants, the researcher may generate descriptions of a culture (Hammersley & Atkinson 1995, p. 67) that may help to focus on the interactions between people and the way they construct contexts. The investigation may reveal patterns of interaction about a group or a culture and explore the life world of individuals. Culture does not just involve specific characteristics of the physical environment, but it also comprehends particular ideologies, values, and ways of thinking of its members (Holloway & Galvin 2016, p. 12). In order to capture these elements connected to the study objectives, there was a careful examination of institutions and courses to understand these particular socio-environmental-educational Arctic contexts.

At first, it was important to define the institutions pertaining to the most representative locations for the study, previously scored according to the criteria explained in Section 1.4 to describe, in the sequence, the universities' department structures where energy courses are offered in Arctic Norway, Canada, and Alaska respectively at the Glacial University, Borealis University, Juniper University and, finally, at the Ice University. It was important to outline the structure of the energy courses offered in the selected universities, to assess their adherence to ESD, to describe the structure of the energy curriculum, and to analyze the processes through which these curricula are 'made' and reviewed.

1.5.1.3 Lecturers

The first group selected was made up of HE lecturers involved directly in undergraduate energy-related modules in the three most pertinent Arctic areas selected as case studies for the research, i.e. Norway, Canada, and Alaska. The interviews provided the opportunity to explore the extent to which their practices adhered to ESD and particularities of the energy curricula under their perspectives.

The process of interviewing participants was when the analytic task started because the researcher had to understand and interpret what the participant was trying to communicate. Transcribing and summarizing interviews was also an analytical exercise followed by a more focused and systematic analysis of the material according to an 'analytic induction' (Pascale 2011, p. 53) meaning a systematic examination of contrasts and similarities across cases to develop ideas and theories.

The qualitative semi-structured interviews were undertaken covering themes of traditional and indigenous education. The interviews were carried on in a receptive style and allowed contrasts to be made concerning the energy component in undergraduate energy courses in these four educational institutions that represent the most advanced programmes and curricula related to energy/SD in these elected locations in the Arctic. These institutions are the most representative in these areas elected by those three levels of criteria previously explained and as a consequence their public is mixed – non-indigenous and indigenous – lecturers and students.

1.5.1.4 Students

The second group was composed of six students from each institution. The students attending the last year of the undergraduate energy-related programmes or, students already graduated were invited via social media (online blog) to voluntarily participate in the study, and they contacted the researcher via email confirming their interest and setting the date, time, and place at the campus for the interviews. Semi-structured qualitative research interviews were carried out according to the interview schedule.

1.6 Phase 3: analysis of data

In the first phase, a documentary research has been developed with a focus on energy education and curricula in Norway, Canada, and Alaska with the aim at investigating the current picture of the energy curriculum in these sensitive areas. The main technique consisted in developing analysis of courses documents to seek information on the courses, programmes, and curricula practised in these locations by making use of library-based and policy-focused documents obtained from specialist government research institutions, and the energy departments of the main universities selected for the study. The process of documentary analysis involved, at first, a superficial examination or skimming, then a comprehensive reading and interpretation. The process combined content analysis and thematic analysis, by organizing information into categories related to the central questions of the study (Bowen 2009, p. 32) and searching for pattern recognition within the data, with emerging themes becoming the categories for analysis (Fereday & Muir-Cochrane 2006).

Given the multicultural characteristics of the study, another relevant feature of the research is the holistic-inductive comparative design. It was adopted to allow openness to the diversity of elements that can emerge from the data (Patton 1990, p. 35) collected from the extensive literature review, case studies, and electronic data on curricula and from the interviews conducted in universities from Norway, Canada, and Alaska that were analyzed as viable and secure geographical areas for the study.

A holistic-inductive comparative design consists in the use of comparative method considering the intercultural comparison and cross-disciplinary work. Comparative method is concerned with systems and functions, rather than structures and rules. It makes use of a functional approach concentrating more on the real-life problems of the societies involved in the study. Macro- and micro-comparison shaded into one another being complementary in the task of systemic study of contextual problems. The comparatist needs to have the ability to communicate, to be persuasive in international contexts (Stein 1977), and at the same time he/she builds trust among international players. The comparatist focuses on the systems in order to provide the framing, the interpretation, and the application of the cross-cultural data.

Apart from being an indispensable method in social theory, the use of comparative analysis can indicate the inadequacies of our usual way of explaining things and open space to new theories about society. Comparison helped to understand the dynamics of social change by providing important clues about the structure and location of power within the given educational context. This allowed contrasts to be made between indigenous and non-indigenous Higher Education students' perspectives, and academics' pedagogical approaches and curriculum design. This analysis also identified similarities and differences among these elements in relation to energy literacy according to intercultural learning experiences and teaching strategies.

Comparison is a fundamental tool of analysis, as it sharpens our power of description and plays a central role in concept formation by bringing into focus suggested similarities and contrasts among cases. Comparison can also contribute to the inductive discovery of new hypotheses and theory building (Collier 1993, p. 105).

Official documents from the studied locations were comparatively analyzed according to document analysis method that provides rich descriptions of actual practices and issues in ESD compared with the stated goals of the respective curricula. The local documents allowed the researcher to evaluate the targeted undergraduate programmes' adherence to UNESCO-ESD paradigm and the needs for innovation in energy education, pedagogies and curricula.

Official documents are likely to be partial or superficial, representing aspirations rather than realities (Shaw et al. 2006). The documents were analyzed considering the degree of alignment of the learning outcomes to the UNESCO-ESD paradigm and the alignment to the pillars of sustainability (social, economic, environmental). The comparative analysis enabled the researcher to evaluate the (un)balance and to identify the main sustainability pillar emphasized in the programme and how this alignment is reflected in pedagogies and curricula. By identifying similarities and contrasts among the documents of the different areas studied the researcher was able to, fundamentally, measure their adherence or not with ESD. The second important element analyzed was the need to include or to improve energy literacy in the curricula and how to do that under different perspectives – lecturers and students, aboriginals and non-aboriginals.

22 Introduction

Document analysis was used in combination with interviews as a means of bringing on board multiple sources of evidence and seek convergence through the use of different sources of data and allowing triangulation (Denzin 1970, p. 291). It also generated useful elements to compose the interview questions and provided a means of identifying and comparing aspects of change and development.

In phase 2, after concluding the interviews transcriptions of academics, a more interpretive work was carried out by segmenting the data created by the qualitative coding to identify patterns of similarity and lines of relationship among memorable statements from the interviews. The same process was carried out for interviews with students. This process generated codes related to the students' experiences and related to their specific perspectives, perceptions, memories, and representations.

The research raised new elements that amplified an emerging understanding of Arctic energy and education. These new elements are the result of an on-going process due to the dynamic nature of the Arctic change and the level of interactions among the stakeholders.

The researcher adopted triangulation respondent validation and constant comparison (Creswell 1998; Lincoln & Denzin 1994) as techniques of validity as a means to treat the data generated from phase 1 and phase 2 as a whole avoiding data fragmentation.

For James (1977, p. 117), 'what really exists is not things made but things in the making'. Similarly, from the epistemological perspective of validation, knowledge is a process of making, instead of a final product. To know means that we acknowledge our relatedness – we are actively connected as knowers to others (e.g., individuals, texts, contexts, and cultures) and to various ways of knowing, including the spiritual, material, and natural worlds (Thayer-Bacon 2003).

The interviews allowed the internal triangulation (see Figure 1.2) of several aspects of lecturers' perspectives, aims, and practices. Moreover, the use of interview schedules with lecturers and students allowed a degree of

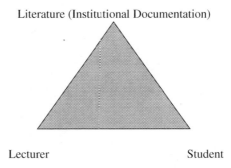

Figure 1.2 Internal triangulation

comparability between staff and students' perspectives, a comparison of the learning outcomes of the four case study institutions and the respective institutional practices, consequently, a comparison of curricula.

1.7 Outline of the book

The book comprises ten chapters. **Chapter 1** has placed the research in the broad context of energy systems. It presented the research rationale and the study focus; it outlined the contribution and its relevance. In this chapter, relevant explanation about methodological approach used to collect and analyze data were conveyed. **Chapter 2** provides a socio-environmental overview of energy under the lens of Education for Sustainable Development. It addresses the complexities of energy production and use and cultural perspectives of a sustainable energy literacy in the Arctic. The chapter also addresses the energy transition and how it affects Arctic societies, the importance of adaptation, and education as a powerful adaptive tool in a changing Arctic. **Chapter 3** presents the theoretical foundation of the study. It presents the theoretical perspective through discussing Complexity Theory, Capability Theory, Educational Theory of Global Citizenship, and Sustainable Development Theory. The conceptual framework also highlights Sustainable Development Theory grounded in the Triple Bottom Line, addressing its three dimensions of performance: social, environmental, and economic. **Chapter 4** presents the findings on the nature of the energy curriculum by outlining the data collected from academic staff interviews, uncovering important components concerning the objectives of the courses and the nature of the energy curriculum in the case study institutions. **Chapter 5** presents the findings focusing on the structure of the energy curriculum in the case studies. It outlines the most influencing factors and distinctions in terms of structure, content, and approaches among the case study courses, showing important characteristics of the curricula in the studied locations. **Chapter 6** presents the findings outlining curriculum design and review according to principles, focus (academic/vocational), and processes adopted in the case study universities. It also highlights elements of curriculum improvement in the selected case studies. **Chapter 7** presents findings based on documents related to course structure, allowing an evaluation of levels of energy literacy and course structure alignment and adherence to Education for Sustainable Development (ESD). This chapter co-relates the documentary findings to previous interview findings, discussing the components of a new model of ESD based on the concept of global citizenship. **Chapter 8** outlines the findings from interviews conducted with indigenous and non-indigenous students and their respective perspectives. **Chapter 9** outlines the pedagogies and approaches to Arctic Higher Education and the relevance of the UNSDGs to the Arctic Higher Education. **Chapter 10** provides an analysis, evaluation and discussion of the research findings with respect to the multidisciplinary and multicultural perspectives of a sustainable energy system

Note

1 Low Arctic designates the part of the Arctic lying between the high Arctic and the subarctic, whose southern boundary lies approximately at the northern limit of tree growth while high arctic designates the northernmost part of the Arctic, including the circumpolar Arctic Ocean with its surface ice and its most northerly coastal margins and islands. The subarctic areas relate to the region immediately south of the Arctic Circle (Ecosystem Classification Group 2013, Ecological Regions of the Northwest Territories – Northern Arctic. Department of Environment and Natural Resources, Government of the Northwest Territories, Yellowknife, NT, Canada, k p. 21).

References

AHDR-HH 2014, *Arctic human development report II*, Norden, Copenhagen.

Aikins, EKW 2014, 'The relationship between sustainable development and resource use from a geographic perspective', *Natural Resources Forum*, vol. 38, no. 4, pp. 261–269.

Alley, RB 2003, 'Palaeoclimatic insights into future climate challenges', *Philosophical Transactions of the Royal Society of London A: Mathematical Physical and Engineering Sciences*, vol. 361, pp. 1831–1848.

Arce, A & Long, N 2000, *Anthropology, development and modernities: exploring discourses, counter-tendencies and violence*, Routledge, London, p. 1.

Arruda, GM 2015, 'Arctic governance regime: the last frontier for hydrocarbons exploitation', *International Journal of Law and Management*, vol. 57, no. 5, pp. 498–521.

Arruda, GM 2018a, *Renewable energy for the Arctic: new perspectives*, Routledge, Abingdon, p. 115.

Arruda, GM 2018b, 'Artic resource development. A sustainable prosperity project of co-management', in GM Arruda (ed.), *Renewable energy for the Arctic: new perspectives*, Routledge, Abingdon, pp. 112, 115.

Arruda, GM & Krutkowski, S 2017, 'Social impacts of climate change and resource development in the Arctic: implications for Arctic governance', *Journal of Enterprising Communities: People and Places in the Global Economy*, vol. 11, no. 2, pp. 277–288.

Baxter, J 2016, 'Case studies in qualitative research', in I Hay (ed.), *Qualitative research methods in human geography*, 4th edn, OUP, Ontario, p. 130.

Berkman, PA & Young, OR 2009, 'Governance and environmental change in the Arctic', *Science*, vol. 324, no. 5925, pp. 339–340.

Bowen, GA 2009, 'Document analysis as a qualitative research method', *Qualitative Research Journal*, vol. 9, no. 2, pp. 27–40.

Boyle, G 2012, *Renewable energy: power for a sustainable future*, 3rd edn, Oxford University Press and Open University, Oxford.

Bridge, G, Bouzarovski, S, Bradshaw, M & Eyre, N 2013, 'Geographies of energy transition: space, place and the low-carbon economy', *Energy Policy*, vol. 53, pp. 331–340.

Brinkmann, S 2013, *Qualitative interviewing. Understanding qualitative research*, OUP, Oxford, p. 37.

Collier, D 1993, 'The comparative method', in AW Finifter (ed.), *Political science: the state of discipline II*, American Political Science Association, Washington, DC. ISBN-13 9781878147080, p. 105, https://ssrn.com/abstract=1540884 (accessed 22 May 2016).

Cotton, D, Bailey, I, Warren, M & Bissell, S 2009, 'Revolutions and second-best solutions: education for sustainable development in higher education', *Studies in Higher Education*, vol. 34, no. 7, pp. 719–733.

Creswell, JW 1998, *Qualitative inquiry and research design: choosing among five traditions*, Sage, London.

Denzin, NK 1970, *The research act: a theoretical introduction to sociological methods*, Aldine, New York, p. 291.

Dethloff, K, Rinke, A, Lynch, A, Dorn, W, Saha, S & Handorf, D 2012, 'Arctic regional climate models', in P Lemke & H Jacobi (eds.), *Arctic climate change*, Springer, London and New York, p. 326.

Dicken, P 2003, *Global shift: reshaping the global economic map in the 21st century*, Guilford, New York.

Elferink, AGO 2011, 'The regime for Marine Scientific Research in the Arctic: implications of the absense of outer limits of the Continental Shelf beyond 200 Nautical Miles', in S Wasun-Rainer, I Winkelmann & K Tiroch (eds.), *Arctic Science, International Law and Climate Change*. Papers from the International Conference at the German Federal Foreign Office in cooperation with the Ministry of Foreign Affairs of Finland, Berlin, 17/18 March 2011. Max-Planck – Institut für ausländisches öffentliches Recht und Völkerrecht, Springer.

Evans, M, Miller, A, Hutchinson, P & Dingwall, C 2014, 'Decolonizing research practice: indigenous methodologies, aboriginal methods, and knowledge/knowing', in P Levy (ed.), *Oxford handbook of qualitative research*, Oxford University Press, New York, p. 179.

Everett, R, et al. 2012, *Energy systems and sustainability*, Oxford University Press, Oxford, UK, pp. 3, 10, 214, 611.

FAO (Food and Agriculture Organization of the United Nations) 2017, *The future of food and agriculture – trends and challenges*, FAO, Rome, p. 4.

Fereday, J & Muir-Cochrane, E 2006, 'Demonstrating rigor using thematic analysis: a hybrid approach of inductive and deductive coding and theme development', *International Journal of Qualitative Methods*, vol. 5, no. 1, pp. 80–92.

Firth, R 2011, 'The nature of ESD through geography', *Teaching Geography*, vol. 36, no. 1, pp. 14–16, p. 14.

Grindsted, TS 2013, 'From the human-environment towards sustainability – Danish geography and education for sustainable development', *European Journal of Geography*, vol. 4, no. 3, pp. 6–20.

Haigh, M 2014, 'From internationalisation to education for global citizenship: a multi-layered history', *Higher Education Quarterly*, vol. 68, no. 1, pp. 6–27.

Hammersley, M & Atkinson, P 1995, *Ethnography: principles in practice*, 2nd edn, Tavistock, London, p. 67.

Hay, I 2016, *Qualitative research methods in human geography*, 4th edn, Oxford University Press, Ontario, Canada, p. 5.

Helm, D 2007, *The new energy paradigm*, Oxford University Press, Oxford, pp. 20, 21.

Heron, J 1996, *Cooperative inquire. Research into the human condition*, Sage, London.

Hilyard, J 2012, *The oil & gas industry: a nontechnical guide*, Penn Well, Tulsa, OK.

Hoberg, EP, Kutz, SJ, Galbreath, KE & Cook, J 2003, Arctic biodiversity: from discovery to faunal baselines – revealing the history of a dynamic ecosystem. *Journal of Parasitology*, vol. 89, pp. 84–95.

Holloway, I & Galvin, K 2016, *Qualitative research in nursing and healthcare*, 4th edn, Wiley-Blackwell, London, pp. 10, 11, 12.

Huckle, J 2002, 'Reconstructing nature: towards a geographical education for sustainable development', *Geography*, vol. 87, no. 1, pp. 64–72.

IPCC-Intergovernmental Panel on Climate Change 2000, *Emissions scenarios. Special report of the intergovernmental panel on climate change*, N Nakicenovic & R Swart, (eds.), Cambridge University Press, Cambridge.

James, W 1977, *A pluralistic universe*, Harvard University Press, Cambridge, MA, (Original work published 1909), p. 117.

Johnson, J, Cant, G, Peters, E & Howitt, R 2007, 'Creating anti-colonial geographies: embracing indigenous peoples knowledges and rights', *Geographical Research*, vol. 45, no. 2, pp. 117–120.

Kelly, AV 2009, *The curriculum. Theory and practice*, 6th edn, Sage, London, pp. 88, 99.

Lemke, P & Jacobi, H 2012, *Arctic climate change*, Springer, London and New York, p. 350.

Levy 2014, *Oxford handbook of qualitative research*, Oxford University Press, New York, p. 179.

Lincoln, YS & Denzin, NK 1994, 'The fifth moment', in NK Denzin & YS Lincoln (eds.), *Handbook of qualitative research*, Sage, Thousand Oaks, CA, pp. 575–586.

MacKay, DJC 2008, *Sustainable energy – without the hot air*, UIT Cambridge, Cambridge.

Mannheim, B & Tedlock, B 1995, 'Introduction', in B Tedlock & B Mannheim (eds.), *The dialogic emergence of culture*, University of Illinois Press, Urbana, IL, p. 2.

Marsh, CJ 2009, *Key concepts for understanding curriculum*, 4th edn, Routledge, Oxford, p. 7.

Maugeri, L 2010, *Beyond the age of oil: the myths, realities, and future of fossil fuels and their alternatives*, Praeger, Santa Barbara, CA.

Munhall, PL 2007, *Nursing research: a qualitative perspective*, 4th edn, Jones & Bartlett Publishers, Sudbury.

Norton, W & Mercier, M 2016, *Human geography*, 9th edn, OUP Catalogue, Oxford University Press, Oxford.

Nussbaum, M & Sen, A 1993, *The quality of life*, Clarendon Press, Oxford University Press, Oxford, New York, p. 20.

OECD 2001, *Nuclear legislation in OECD and NEA countries, regulatory and institutional framework for nuclear activities*, OECD, Norway, Paris.

OECD 2011, 'Green growth studies: energy'. OECD Publishing, p. 9, www.oecd.org/greengrowth/greening-energy/49157219.pdf.

Parkhe, A 1993, '"Messy" research, methodological predispositions and theory development in international joint ventures', *Academy of Management Review*, vol. 18, pp. 227–268.

Pascale, CM 2011, *Cartographies of knowledge: exploring qualitative epistemologies*, Sage, Thousand Oaks, CA, p. 53.

Patton, MQ 1990, *Qualitative evaluation and research methods*, Sage, Newbury Park, CA, p. 35.

Poppel, B 2015, *SLiCA: Arctic living conditions – living conditions and quality of life among Inuit, Sami and indigenous peoples of Chukotka and the Kola Peninsula*, Nordic Council of Ministers, Denmark, p. 67.

Radcliffe, S 2006, *Culture and development in a globalizing world. Geographies, actors, and paradigms*, Routledge, London, p. 237.

REN21 2017, *Renewables 2017 global status report*, REN21, Paris, France.

Rubin, HJ & Rubin, IS 1995, *Qualitative interviewing: the art of hearing data*, Sage, London, pp. 17, 34.

Schaefer, RT 2017, *Glossary' in sociology: a brief introduction*, 4th edn, originally c. 2000, McGraw-Hill Ryerson, Whitby, Ontario, http://novellaqalive.mhhe.com/sites/0072435569/student_view0/glossary.html, site dated 2017 (accessed 11 June 2018).

Senge, P 1990, *The fifth discipline: the art and practice of the learning organization*, Century Business, London, p. 68.

Shaw, S, Elston, J & Abbott, S 2006, 'Comparative analysis of health policy implementation', *Policy Studies*, vol. 25, no. 4, pp. 259–266.

Stein, E 1977, 'Uses, misuses', *Nonuses of Comparative Law*, vol. 72, no. 2, p. 198.

Sterling, S 2003, 'Whole systems thinking as a basis for paradigm change in education: explorations in the context of sustainability', Unpublished doctoral dissertation, University of Bath, Bath.

Sterling, S, Irving, D, Maiteny, P & Salter, J 2005, 'Linking thinking: new perspectives on thinking and learning for sustainability', in SB Aberfeldy, JD Smith, PM Hildrethand & C Kimble (eds.), *Knowledge networks: innovation through communities of practice*, IGI Publishing, Pennsylvania, PA, pp. 150–164.

Stern, NH 2007, *The economics of climate change: the stern review*, Cambridge University Press, Cambridge, pp. 20, 65, 68.

Strauss, A & Corbin, J 1990, *Basics of qualitative research*, Sage, Newbury Park, CA, p. 61.

Taylor, C 1989, *Sources of the self: the making of the modern identity*, Cambridge University Press, Cambridge, MA, p. 36.

Thayer-Bacon, B 2003, *Relational "(e)pistemologies*, Peter Lang, New York.

Thomas, G 2007, *Education and theory*, Open University Press, London.

Tilbury, D 2011, *Higher education in the world 4. Higher education's commitment to sustainability*, Barcelona Palgrave and Global Universities Network for Innovation (GUNI), p. 3.

UNWCED 1987, 'Report of the world commission on environment and development', General Assembly Resolution 42/187, 11 December 1987, p. 10, www.un.org/documents/ga/res/42/ares42-187.htm (accessed 30 March 2018).

USGS 2008 'Circum Arctic resource appraisal: estimates of undiscovered oil and gas North of the Arctic circle', USGS Fact Sheet 2008–3049.

van Voorst, RS 2009, 'I work all the time- he just waits for the animals to come back: social impacts of climate changes: a Greenlandic case study', *Jàmbá: Journal of Disaster Risk Studies*, vol. 2, no. 3, pp. 235–252.

Waters, T 2007, *The persistence of subsistence agriculture*, Lexington Books, Lanham, MD, pp. 3, 48.

Wilde, V 1992, 'Controversial hypotheses on the relationship between researcher and informant in qualitative research', *Journal of Advanced Nursing*, vol. 17, no. 234–242, p. 240.

2 The Arctic context
Energy, development, education

2.1 Introduction

This chapter provides an overview of energy under a socio-environmental perspective and through the lens of ESD. It addresses the complexities of energy production and use, and the intercultural perspectives important to the development of sustainable energy literacy in the Arctic. It also approaches the concept of Traditional Knowledge (TK) and the importance of HE for indigenous people in the Arctic.

It is important to state that 'energy' is considered here to be a geographical concept and an element of change, as its production/use presents different dynamics in distinct geographical contexts (bio-regions). The literature review concludes by arguing that students will develop greater capabilities to deal with the local energy reality if they acquire geographical knowledge related to the energy scenario in the Arctic.

The text is structured in a way that the reader can be shown the energy scenario in the Arctic, specifically in Norway, Canada, and Alaska, where very relevant energy installations are located. This scenario is a brief picture of an on-going process of energy transition, in which the traditional oil and gas reserves are being explored and the remaining fossil unexplored reserves are considered of good potential by southern nations. In the same Arctic landscape, there are renewable energy projects being initiated; however, a better understanding of the energy-transition process and of sustainable energy resources is required to deal with these different energy scenarios across the Arctic.

The chapter aims to expand the understanding of this scenario and how energy production is linked to well-being and human development in a complex geographical area where cultural aspects are entangled with environmental and social factors by showing that the Triple Bottom Line (TBL) (Elkington 2004, p. 23) cannot be limited to the traditional three pillars of sustainability. The text also brings clarity about energy production/use, impacts and adaptability to transitions, and the relationship between these points in order to propose a reflection on Higher Education (HE), geo-capabilities, knowledge, and skills associated with geographical factors because the reality in Norway is different from Alaska and Canada showing that there are many 'Arctics', instead of only one. The local energy potentials are different, and another variable, TK, started to be considered in certain processes of communication and knowledge creation, locally.

2.2 Energy and the Arctic

This section outlines the different energy resources available in the Arctic as well as the explored and unexplored energy reserves in the region. This spatial energy scenario is useful for understanding the geographical energy production areas that justify the Arctic nations in the study, their production capacity, the specific socio-environmental contexts associated with the levels of energy production/use, and contexts related to the local energy education in the studied Arctic areas.

The development of industrial activities in the Arctic and sub-Arctic regions is not a new venture being part of the modernity process contrasting to traditional ways of living. Over centuries, alongside fishing, shipping, tourism, and mining, oil and gas represented an important industrial activity due to the large fossil fuel resources concentrated in areas of Alaska North Slope, Beaufort Sea, Barents Sea, Amerasian Basin, and East Greenland Rift Basin (Arruda 2018a).

The Arctic is home to an unexplored cluster of natural and human resources accounting for 20% of the world's fossil fuel energy resources such as oil, gas, and coal (USGS 2008) and an estimated 23 billion barrels of oil and 108 trillion cubic feet of natural gas offshore are recoverable from the Outer Continental Shelf (USGS 2008). More than 70% of oil is estimated to lie in just five geologic provinces: Arctic Alaska, The Amerasia Basin, the East Greenland Rift Basins, the East Barents Basins, and the West Greenland – East Canada sector. More than 70% of natural gas is estimated to be in only three provinces: The West Siberian Basin, the East Barents Basins, and Arctic Alaska along the North Slope (Perry & Andersen 2012, p. 14). The Arctic region comprises vast amounts of extractive energy resources, mostly in offshore areas, which have been unreachable for a long time, but the current climate change and associated melting of ice is making these accessible (Figure 2.1). The main areas of extractive potential are the North Slope in Alaska (US), East Siberia, West Siberia, The Timan-Pechora area, South/North Barents (Russia), the area east of Norway, and East Greenland. Other important areas include the Vilkitsky sector, the Laptev Sea, the Vilyuy sector, the Khatanga area and the North Kara Sea (Russia); the West Barents Sea north of Russia; Denmark and Norway; the West Greenland area; the Sverdrup area north of Canada; and the Beaufort-Mackenzie area north of Canada and Alaska.

Yet, the Arctic environment is extremely sensitive. Important conclusions about the challenges ahead were discussed in 'Oil and Gas in the Arctic – Effects and Potential Effects' between 2004 and 2007 under the auspices of the Arctic Monitoring and Assessment Program (AMAP 2007, p. 10).

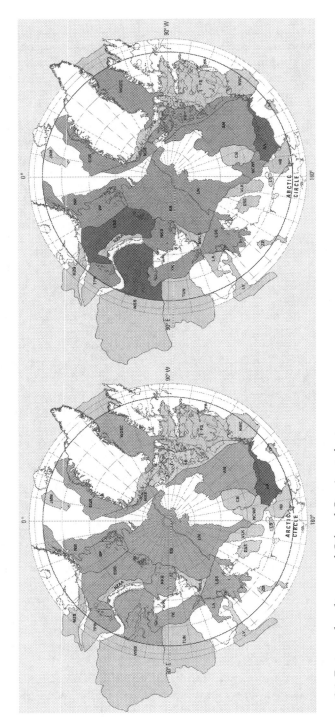

Figure 2.1 Estimated undiscovered Oil and Gas Arctic basins.
Source: U.S. Geological Survey (2008)

Oil and gas potential are significant in Arctic Alaska. On Alaska's North Slope, the Nation's largest oil field, Prudhoe Bay, has been in production for several decades. Oil has been produced from the Beaufort Sea OCS since the early 2000s, and the Arctic OCS potential for production of additional oil and gas resources is significant (AMAP 2007). Beyond petroleum potential, this region supports unique fish and wildlife resources and ecosystems, and indigenous peoples who rely on these resources for subsistence (Arruda 2015). While the potential for, and interest in, energy resources is clear, there is significant public discourse over the ability to develop oil and gas resources safely (Holland-Bartels & Pierce 2011, p. 4), to understand environmental and social consequences of any development (The Nature Conservancy and Wainwright Traditional Council 2008), and to implement effective impact prevention and mitigation strategies (Holland-Bartels & Pierce 2011, p. 4) considering different views on the sufficiency of the scientific information available to well evaluate energy development options and understanding on environmental sensitivities.

Challenges exist in the Arctic fossil fuels exploitation ventures, and are generally linked to technical, environmental, economic, commercial, and social aspects. Remote locations with sensitive environments, problems with ice on equipment and safety design of equipment, expensive operations and impacts on Arctic communities are a few examples of these challenges; however, the eight Arctic nations have an invaluable opportunity to make genuine and effective changes in how the Arctic is viewed and to create new development and energy models for the future Arctic region. At the same time, it is possible to envisage a great potential for economic and energy prosperity, it is undeniable that it is currently a region of high levels of energy poverty, social sensitivity, workforce unpreparedness, and environmental fragility (Arruda 2015, p. 500).

2.3 Energy transition and governance: how it is transforming the Arctic

Historically, energy transitions have been associated with broad social change. This is no different with the 21st century energy challenge that Bridge et al. (2013) believe will bring about a new transition, towards a more sustainable energy system characterized by universal access to energy services, security, and reliability of supply from efficient, low-carbon sources. However, despite significant reserves of oil and gas in the Arctic, there is still great potential for renewables in the region. Indeed, renewable energy sources that have been considered in governmental plans locally, regionally, and internationally (IEA 2013, p. 18) could support the energy transition process. Plans materialized in small projects in several scattered parts of the Arctic (Arruda & Arruda 2018) demonstrate the start of an important energy transition trend, which can potentially transform the Arctic energy matrix, but are unlikely to replace the need for finite energy reserves. In the areas under study in this chapter, i.e. Alaska, Canada, and Norway, there is potential for wind power, solar, and geothermal energy production and use. These will now be considered in order to provide country-specific contexts, important for understanding the nature of the different regions in which the HE courses dealing with energy are situated.

Alaska is a major producer of petroleum resources and it is less known for its significance in renewable energy production. However, in the southeast and southcentral parts of Alaska there are important hydropower projects due to the climatic and geological features of the area. In western Alaska, the wind resources are abundant due to the Chukchi Sea, Bearing Sea, and Pacific Ocean winds (REAP 2016). The solar resource varies across the state and geothermal and heat pumps that use heat from air, ground and ocean are also growing in number (CCHRC 2016; REAP 2016). Emerging technologies are transforming Alaska's energy systems dictating the pace of the Arctic region energy transition. There are investments in hydrokinetic, wave, and tidal energy sources, but the great challenge, apart from the vast distances to integrate, is the skills and local capabilities to make these projects tangible in practice.

To understand the patterns of energy production and use in the Arctic it is important to examine quantitative information related to the local energy matrix. This practical information shows the main energy sources being explored (energy matrix) and how the matrix is being transformed by new sources. The latest information published by the National Energy Technology Laboratory (NETL 2017) gives evidence of the expansion capacity of the industry in these Arctic locations, as well as the present and future educational and professional specialties the energy industry demands. The future energy professionals need to know about the energy matrix model to operate the local energy industry and to understand how to make decisions on sustainable energy resources according to specific geographical energy, environmental, and social contexts.

Currently, Alaska generates 52% of its electric power from natural gas produced in Cook Inlet near Anchorage (EIA/DOE 2017). Hydropower is the next largest contributor, providing 24.9% of electric generation, followed by oil at 13.6%, coal at 6.1%, and wind at 2.5% (EIA/DOE 2017). While natural gas is prevalent in more urban areas, power in most of rural Alaska is generated by diesel fuel (59%), followed by hydro at 29%, natural gas at 8%, and wind at 5% (EIA/DOE 2017). The consequence of this matrix model is that the cost of energy in most Alaskan communities is high, as the cost of heating is high, which makes it difficult to think about sustainability when people are cold and the poorest households spend up to 47% of their income on energy, representing more than five times their urban neighbours (AACHC 2016, p. 14). In the Northwest Territories of Canada, the energy scenario is similar to the Alaskan Arctic, as it is still powered by old diesel-fired power plants. Communities like Nunavut, for example, are completely reliant on diesel fuel, and in the Northwest Territories diesel represents the most expensive source of energy, which makes energy transition a real concern for the decarbonization agenda. Northerners use about twice as much energy as the national average, due to the long, cold, and dark winters they endure. The implication of being reliant on diesel is the large amount of greenhouse gases emitted in this process as the fuel is shipped over long distances as well as causing local pollution. There has been limited progress in developing renewable energy sources that could reduce these communities' reliance on diesel because from 80 communities in Canada's territories, 53 are still dependent on diesel fuel (Senate of Canada 2015, p. 9).

In Norway, 98% of the electricity production comes from renewable energy sources, hydropower being the main vector of the Norwegian production. Hydropower has been the basis for Norwegian industry and the development of a welfare society since Norwegians started utilizing the energy in rivers and waterfalls to produce energy in the late 1800s. Since then, Norwegian hydropower has become an increasingly important part of Norwegian society. The usage of electricity has increased in line with the modernization and economic growth in Norway projecting the country as an electricity producer rather than a traditional oil and gas producer, demonstrating that sustainable energy is viable in the Arctic (Prime Minister's Office 2015, p. 13).

Geographically speaking, approximately half of the communities residing within the Arctic are not connected to a traditional power grid, but they are, instead, dependent on remote obsolete micro-grids to provide electric power services and this fact increases the complexity and the amount of investment needed to integrate renewables in the Arctic. The knowledge of contextual energy scenarios in the studied locations reveals that old systems of energy production like diesel power plants co-exist with transitional energy sources like natural gas that is presently an available attempt to replace the obsolete energy system, at the same time that renewable sources started to be a tangible possibility but requiring infrastructure. For energy education in the Arctic, the complexity resides not only in the external factors like socio-environmental effects, but also in the energy matrix transition and how it is transforming the Arctic.

2.3.1 The influence of political decision-making on the Arctic transformation

Fossil fuel production and use is not a new venture in the Arctic areas rich in oil and gas. In fact, oil and gas activities are contributors to regional and national economies due to revenues from taxes, royalties, exports stimulating local and national businesses, education, and development. In general, the fossil fuel industry has been a major driver for the development of local infrastructure, employment, and industrial development (Berkman & Vylegzhanin 2010, p. 182) besides being a national security issue. The types of energy exploration in different geographical areas reveal the characteristics of the energy matrix and the local energy system with implications for governance, political decision-making, and educational systems. In Norway, the energy system is under the central government jurisdiction, and it is currently driven by the local use of clean energy generated by the smart grids integrating different modalities of renewable energy (solar, wind, tidal), but the country adopted a policy of producing fossil fuels that are exported to third countries. These geographical differences in terms of energy production, use, and investment have important influence on the local educational systems.

Alaska, for centuries, has been a traditional fossil fuel producer, but the price regime of fossils for the locals has triggered a reverse process of adopting renewable energy mainly for off-grid communities that cannot afford a living style based on imported diesel fuel. This scenario is re-orienting the political spectrum at local and regional levels, but it is still in need of a national energy policy that integrates renewables into the energy matrix, apart from the local renewable energy initiatives developed by educational institutions and funded by the private sector. The current central government administration and its severe political constraints to HE investment tend to intensify the private sector investments in Alaskan HE.

Canada presents a diverse governance scenario. The country is one of the largest producers of energy in the world, and is well known for being the fourth largest natural gas exporter and the sixth largest oil producer with net exports from crude oil accounting for approximately 30% of Canada's total annual exports or revenues estimated at $97 billion (Natural Resources Canada 2018). In Canada, provincial governments have shared jurisdiction over energy and authority over exploration, development, management, and trade of energy production (Constitution Act 1867). Canadian provincial governments also have jurisdiction on educational systems and important funding comes from the energy industry to develop the education sector. As Canada has no federal department of education, departments established at the provincial level are responsible for overseeing the administration of HE. The lack of a unified, integrated national system results in regional differences in terms of the way educational standards and curricula are determined.

Canada continues to be dominated by fossil fuels (oil and gas, shale gas) production and use due to the significant exports to the U.S. and renewable energy does not have the prominence that it has in the neighbour Alaska, where scientific initiatives are influencing and re-orienting public policies and HE to renewable energy. The co-existence of fossil production and renewable energy projects tend to dominate the political spectrum and decision-making for the next 10 years due to the increasing global energy consumption that might surpass the capacity of energy markets and because of the new areas been unveiled by the ice melting.

The basic functions of the process of Arctic governance should involve activities related to goal selection, coordination, implementation, and accountability in terms of energy decision-making and energy education. It should imply designing the highest levels of responsible and durable development based on energy literacy. Goal selection and coordination require acknowledgement and integration of development goals across all levels of the system by establishing priorities at different levels and sectors. Implementation can be performed by state actors but also may involve social actors (Levi-Faur 2012) and educational institutions.

Additional scientific understanding of biophysical and socio-economic systems in the Artic is also necessary to formulate adequate policies, legislation, and enforcement mechanisms, but an extremely important step is to strengthen interdisciplinary research and knowledge in the region through international educational cooperation agreements and diplomacy. This is a unique opportunity to expand and integrate science and diplomacy (Berkman & Young 2009). The political scenario points to a permanent shift that will clearly affect the energy systems of the future with investments shifting from fossil fuels to renewable sources of energy.

Governance in the 21st century Arctic is assuming new contours to consider the energy transition process and the severe impacts of climate change in reshaping policies and HE. This transition should have as a base, more inclusive, democratic, and durable patterns of decision-making to reshape the current governance model from a state-centred to a more social-centred governance style (Arruda 2015, p. 504). The new governance system must be an instrument of equality and efficiency more than an instrument of technological control on natural resources and must recognize the gaps between the formal constitutional order and the way it is operated in reality (Arruda & Krutkowski 2017, p. 284).

2.4 How energy production and use affect Arctic societies

After examining the main characteristics of the energy matrix in Norway, Alaska, and Canada, it is important to look at the multidisciplinary and multicultural aspects of energy production and use by understanding how energy affects societies. This section discusses the human geographical contexts related to energy that represent another perspective to be considered when educating future energy professionals as they will need to understand the connections between energy-induced impacts, climate change, and Sustainable Development.

The Arctic is the home of a population of approximately four million people scattered among provinces of Alaskan, Canadian, Scandinavian, and Russian Arctic. Arctic indigenous peoples have lived in Arctic lands for thousands of years and relied on hunting, fishing and gathering to sustain their communities. Their traditional values, activities, and practices contributed to the development of unique economies and community lifestyles. Apart from being a region full of potential for sustainable growth, there are significant uncertainties associated to global economic interests, to local complexities in relation to climate change and energy access that are still challenging governments.

Fossil fuels are believed to be the primary cause of climate change and it has global and local impacts due to the human-induced atmospheric build-up of greenhouse gases (GHGs) (Nanda & Pring 2013, p. 284). The burning of coal, oil, and natural gas in developed countries accounts for the overwhelming majority of human-caused emissions of carbon dioxide (CO_2). Stern (2007, pp. 65, 68) explains in detail how climate change will affect people around the world in *The Economics of Climate Change*. He emphasizes that climate change threatens the basic elements of life compromising access to water, food, health and use of land and the environment. He states that the consequences of the phenomenon will become disproportionately more damaging with increasing warming that will trigger the melting of glaciers, rising sea levels, ocean acidity, black carbon (a type of carbonaceous material with a unique combination of physical properties), declining of crop yields, permanent displacement and severe biodiversity loss, among other impacts (Bond et al. 2013, p. 5380). Meehan (1995, p. 519) points out the diversity of ecological pressures faced by Arctic ecosystems and societies as a result of anthropogenic activities, among them those associated with oil and gas exploration and mining and smelting of heavy metals (Rees & Williams 1995, p. 253). Forbes (1995, p. 372) complements this stating that local and regional impacts are of two orders: 'acute disturbances' resulting from a single disruptive event like the impacts of vehicles passing across tundra landscapes or a temporary oil exploration rig; and 'chronic disturbances' resulting from a sustained long-term pressure upon the ecosystems like the deposition of pollutants, heavy metals, or permanent oil exploitation sites representing permanent or semi-permanent environmental changes. The scale of the changes is proportional to the local industrialization level as the energy-induced impacts are relatively small-scale in the Canadian Arctic in comparison to the northwest Siberia where the effects are persistent.

The Arctic is affected by energy produced and used in the Arctic and in the non-Arctic world, as climate change is the common point of these two geographical spaces. Arctic and sub-Arctic waters are particularly prone to ocean acidification compromising marine animals, base of the food chain for ecosystems and communities with severe impacts on fisheries. Another severe Arctic problem is the deposition of black carbon or soot, which absorbs the sun's heat and warms the atmosphere, thus melting the snow and ice beneath it (Mathis et al. 2015, p. 72; Stohl et al. 2013, p. 8834).

In terms of local energy production and use, the Scandinavian Arctic is the most developed part of the Arctic region. There are more cities, universities, cultural institutions, and economic development there due to the offshore oil and gas resources and plentiful hydropower resources that have long been the energy matrix sustaining Norway's economic success. Finland, Sweden, and Norway are part of the common Nordic and Baltic wholesale electricity market, which uses Nord Pool Spot of Norway as its trading centre. This is possible because of the interconnectivity of these countries' electricity systems, ensuring sufficient transmission capacity in a common electricity grid. Norway's prime minister's office believes that the production of renewable energy through green certificates can lead the country towards a low-carbon green economy, helping to overcome the use of natural gas (LNG) that, despite having the lowest CO2 emissions per unit of energy, still represents the fossil fuel presence in several essential mineral and petrochemical processing operations locally (Prime Minister's Office 2015, pp. 14, 15).

In the Canadian Arctic, the scenario is a bit different as societies have been reliant on diesel-based electricity generation for long time. The result of this dependency has a range of impacts related to black carbon, the risk of diesel spill during the transportation, and the region's dependence on costly diesel generated electricity. Communities in the territories, most of which are indigenous, are predominantly small, isolated, and widely dispersed across an immense landmass. These communities are considered off-grid since they are not connected to the North American electricity or natural gas grids. While hydropower is available for many communities, northerners rely on carbon-intensive fuels such as diesel much more heavily than the rest of Canada. Diesel is often the only reliable option for heat and electricity in the Canadian Arctic despite the high costs of transportation and environmental disadvantages (Senate of Canada 2014, p. 9). The Arctic context is very specific and complex in physical, human, and industrial levels, for this reason it is beneficial to promote sustainable energy literacy in the region to broaden understanding on the potential and the benefits of low-carbon energy sources.

In Alaska and in Canada, hydropower resources provide electricity for the larger population centres; however, the northern communities are still reliant on diesel. Canada and Alaska share similar challenges and resources; however, these countries present different institutional structures: the reason why Alaska has developed more renewable resources than Canada. In the Alaskan Arctic, every community without exception has access to some type of renewable energy resource. The great challenge has been to create the adequate infrastructure to integrate renewable energy technology into a grid structure and create an energy-efficient system. These are challenges that depend on energy-literate communities from both the urban centres and remote areas.

Successful deployment of initial renewable energy projects in the Arctic demonstrated the viability of alternative sources and it will require active community participation in energy projects. According to Marsik and Wiltse (2018, p. 13):

> There are many factors that need to be considered when evaluating renewable energy sources for the Arctic. Important factors to consider include: magnitude of resource, proximity of the resource to the point of use or existing distribution lines, land ownership and public use issues, permitting, maturity of available technologies, suitability of technologies for cold temperatures, aesthetics, longevity of technologies,

skills of local workforce to operate and maintain installed systems, life-cycle economics and environmental impacts, effects on grid stability, risks of catastrophic failures and other safety issues, acceptance of the project by local culture, and other factors.

(Marsik & Wiltse 2018, p. 13)

The active social participation in the local energy debate depends on efficient energy-literacy initiatives. Alaska, under the leadership of the Arctic Council, is investing in a fundamental change of perception of the feasibility of renewable energy projects in the Arctic (AACHC 2016). Due to the work of this high-level intergovernmental forum responsible for enhancing cooperation, coordination, and interaction among the Arctic States, local communities are being incentivized to engage in the sustainable production and consumption of energy, which will consequently make them turn to renewable energy resources and keep diesel as a backup energy source. The communities in the urban centres and more remote areas have been informed and engaged in the impacts of energy production as part of independent communication strategies but a more systematic approach is required. This change in perception in relation to energy will reduce the production of black carbon as well as the risks of fuel spillage into the pristine Arctic environment.

2.5 The importance of adaptation for Arctic societies in times of energy transition

The speed of changes and energy insecurity makes it hard to predict the reality that Arctic societies will face from 10 to 50 years into the future. Currently, it is not possible to determine future vulnerable spots, or the scale of outcomes generated by combination of climate change, energy demand, pollution, and human activity. It is also an *incognita*, how this set of risks will affect the ecosystem's ability to adapt to dynamic conditions. This level of complexity and unpredictability indicates the need to reflect on understandings of energy and development in changing contexts. This is what is at stake in this study of HE curricula and pedagogies.

According to Blaauw (2013, p. 180), fossil fuels meet 80% of the global energy demand, but increasingly the energy mix must be supplemented by a diverse range of energy sources, from fossil fuels (gas) to renewable sources, taking approximately 30 years for a new energy technology to reach 1% share of the energy mix. He adds that by 2050 these energy sources are expected to still meet 60% of demand and most part of these resources have not been found or will be more expensive to produce making the energy resources from the Arctic Ocean an important future source.

For Smit and Wandel (2006, p. 283), under the auspices of the Framework Convention on Climate Change (UNFCCC), vulnerability refers to the manner and degree to which a community is susceptible to conditions that directly or indirectly affect the well-being or sustainability of the community. Adger and Kelly (1999, p. 254) believe that ecosystems are included in this concept and vulnerability is a function of exposure, sensitivity, and adaptive capacity. Exposure and sensitivity refer to the manner and degree to which a community is susceptible to specific conditions of stress and hazards at a specific time and location. The way the community responds to these stimuli depends on their adaptive capacity, conceptualized by Berkes and Jolly (2002, p. 18) as 'the ways in which individuals, households, and communities change their productive activities and modify local rules and institutions to secure livelihoods'.

In terms of the Arctic societies targeted by this study, their specific geographical position, energy production and use methods and human development features bring different implications of economic, environmental, and social importance for their adaptation methods and capabilities. Coastal and in-land Arctic communities have specific adaptive strategies to deal with challenges related to ice/snow anomalies, extractive industry, and climatic variability. These factors reinforce the premise that energy learning and literacy should be performed under multidisciplinary and multicultural perspectives in order to promote a broader understanding of the scale of impacts caused by the energy produced in Arctic and non-Arctic nations.

In settlements of the Canadian Arctic like Tuktoyaktuk and Ulukhaktok (Figure 2.2), located in the Inuvialuit Region, 57% of community members of the former and 76% of the latter participate in hunting and fishing, reflecting that traditional activities of food production and consumption are a fundamental aspect of Inuit culture and food security (Andrachuk & Pearce 2010, p. 64).

In both locations, shown in Figure 2.2, environmental changes related to the land and sea ice dynamics have been observed (Andrachuk & Pearce 2010, p. 65). Out-of-season patterns of wind and precipitation contributed to degrade the wildlife habitats altering harvesting seasons (Andrachuk & Pearce 2010, p. 68). Coastal erosion is an additional risk that has been affecting the local infrastructure determining measures of shoreline protection (Andrachuk & Pearce 2010, p. 72). In both, Ulukhaktok and Tuktoyaktuk, adaptability is evident in strategies being employed to deal with current climate related exposure-sensitivities but the feasibility of many adaptive strategies depends on non-climatic factors like investments, the ability to secure consistent employment and a steady source of income in order to purchase boats, snow machines, firearms, and ammunition (Hovelsrud & Smit 2010, pp. 69, 71).

Hovelsrud and Smit (2010, p. 75) also emphasize the important aspect of Inuit lifestyle as being the capacity to deal with variable environmental conditions due to their Traditional Environmental Knowledge (TEK) and land skills generated through hands-on experience that has been transmitted through generations. Another important field observation is that the Inuit younger generation is spending considerably less time in subsistence activities outside of land camps and there is a real concern about the risks the young Inuit can face due to the changing environmental conditions as the

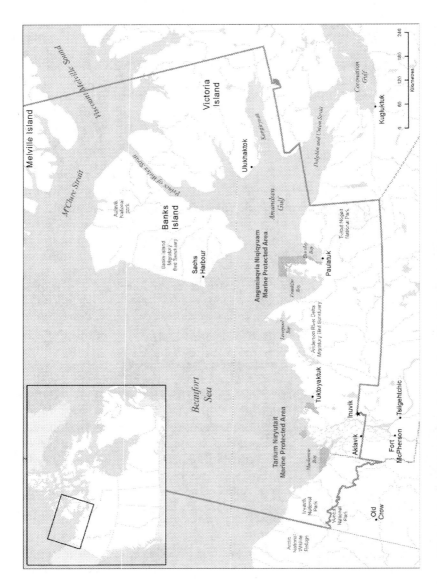

Figure 2.2 Location of Tuktoyaktuk and Ulukhaktok in the Inuvialuit Settlement Region, NWT, Canada. Source: University of Guelph, Global Environmental Change Group, Jan 2010

indigenous youth are not well prepared for them (Bitz et al. 2016). For this reason, it seems important to encourage initiatives to develop adaptive knowledge and skills. These learning initiatives become even more important when observations in these two locations pointed out to the growing population of Inuit youth and the decline of the elder population responsible for the knowledge transmission in these communities. The learning factor and the availability of funding are two central aspects affecting the adaptive capacity of these two communities concerning the risks of environmental change.

In the Norwegian context, the region relies on natural resources as an essential part of the economy, culture, and lifestyle (Norwegian Government 2017). The relatively mild climate provided by the Atlantic Ocean influence and the ocean temperatures in the Barents and Norwegian Seas provided the adequate conditions for human settlements that reveal a combination of fishing and farming activities (reindeer herding, aquaculture) and public sector employment reflecting the modern application of technologies and the great concern for fishing policies and regulation (Duhaime & Caron 2017).

Generations of fishermen and reindeer herders, for more than 1000 years, developed around these activities, their communities' economies, and lifestyle by using their TK to manage their local resources and supply their needs at different levels (Nordic Council of Ministers 2015). These communities have been subject to climatic variability as the ocean temperatures increased, extreme weather events are frequent, winter precipitation is abnormal and sea level rise could be observed in specific areas according to the latest data collected locally by scientific missions like CAVIAR (Community Adaptation and Vulnerability in Arctic Regions) (Hovelsrud & Smit 2010, p. 3) and SLiCA (Survey of Living Conditions in the Arctic) (Poppel 2015, p. 67).

The collaborative effort of using traditional and scientific knowledge can certainly reduce vulnerability and, at the same time, strengthen Arctic communities' resilience and support-capacity to implement, in the field, stronger co-management adaptive strategies (Arruda 2018b) in response to climate change (Arruda & Arruda 2018). A joint assessment and management on impacts issues has been performed, in recent years, based on information exchange, and cooperation grounded on co-management models of data collection, analysis, communication, and implementation at the local level (Arruda & Krutkowski 2016, p. 519).

2.6 Education as a powerful adaptation tool in a changing Arctic

The approaches to Arctic HE seem to be in need of a more refined connection to energy contexts than never before. Despite the local energy contextual differences in the areas selected for this study, the common factor seems to be the severe environmental changes triggering a range of societal impacts in need of analysis and proactive solutions. Who will provide the analysis and solutions it will depend on the level of awareness and literacy oriented to resilience and adaptation.

Arctic HE plays a key role in the local long-term community's adaptation strategies (Arruda 2018a). As the ice retreats, the current challenges concerning the environment, and human extractive activities, will intensify even more the effects of climate change on Arctic ecosystems and communities. Climate change and modernization have thus become two intrinsically linked forces that severely alter the context in which the indigenous populations of the region sustain a livelihood (van Voorst 2009, p. 236). Local animal and plant species are of dietary importance, while hunting, fishing, or foraging are all of cultural and social value. Modernization, as in the Western societies, implies an energy-intensive commercial economic system distinct of the Arctic original economy and development patterns.

Adaptive strategies are the ways in which individuals, households, and communities change their productive activities and modify local rules and institutions to secure livelihoods (Berkes & Jolly 2002, p. 18). The flexibility of this more modern and resilient way of thinking and operating, in practice, allows communities to deal with climatic uncertainty and constantly adapt to change, in real time, by using information systems to disseminate their information within their communities.

Reasonable levels of adaptation to a changing Arctic seem to depend on efficient communication and knowledge creation (NTI 2005). The collaborative effort of using traditional and scientific knowledge/data can certainly reduce vulnerability and, at the same time, strengthen Arctic communities' resilience and support capacity to implement, in the field, stronger co-management adaptive strategies (Arruda 2018b) in response to climate change and the off-sets of energy production.

TK is a fundamental part of traditional ways of living, learning, and traditional community expertise or *birgejupmi (livelihood, survival capacity)* (UNPFII9 2010), *isumatturuk* and *annaumanasungniq (activities and the capability to make a sustainable livelihood)* (Webster & Zibell 1970). These terms relates to the individual and collective capacity of indigenous communities to maintain themselves in an area with its respective resources by employing traditional know-how, skills, critical-thinking, social competence by integrating communities, landscape, natural environment, ecosystems, social and spiritual development and identity, but also the past, the present, and the future (Government of Nunavut 2007, p. 23; Vars 2007).

According to Martínez-Cobo (1982, p. 3) indigenous communities, peoples, and nations are those which, having a historical continuity (ancestral land occupation, culture, and language) with pre-colonial societies, develop, preserve, and transmit to future generations their ancestral territories, TK, and identity according to cultural patterns, social, and legal institutions. TK is distinct of other knowledges because it is often oral, invisible, and difficult to measure (tacit), as it is part of the indigenous communities' wisdom personified by the indigenous knowledge keepers. TK is conceptualized as the knowledge and know-how unique to a given society or culture encompassing 'the cultural traditions, values, beliefs and worldviews of local people' (Agrawal

1995, p. 418) vital for the historical and cultural continuation of a particular group (Magni 2016, p. 5). Based on its practicality and dynamicity (Sillitoe 1998), it is a knowledge in constant evolution through creative thinking guided by strong contextual, pragmatic, utilitarian features of the everyday demands of life (Briggs 2005, p. 10) that can offer interesting strategies to cope with global challenges.

Joint assessments on impacts issues have been performed, in recent years, based on information exchange between scientists and indigenous populations, on experience-based data and scientific data. There have been important joint initiatives like EALÁT (Reindeer Herders Vulnerability Network Study), initiated in 2006, in Norway, by the Association of World Reindeer Herders (WRH) and the Arctic Council Sustainable Development Working Group (SDWG) focusing on reindeer herding, TK, adaptation to climate change and loss of grazing land to the extractive industry.

Another important example of how Arctic communities are being benefited by the use of specific educational approaches of combined scientific and TK is the Inuit Local Observations and Knowledge of the Arctic launched in 2007 (ELOKA) (McCann 2013), consisting in a management research support service that specializes in working with Arctic communities and researchers in the collection, preservation, and use of local and TK and community-based monitoring data. In association with the University of Colorado's National Snow and Ice Data Center, the project focuses on creating datasets ranging from qualitative data such as map overlays, transcripts, and audio/videotape interviews of subsistence hunters who hold valuable knowledge on sea ice, to quantitative data such as snow and ice thickness, temperature, and wind velocity (Pulsifer et al. 2012, p. 272).

The indigenous system of conceptualization, classification, and analysis comprehends the application of physical and strategic knowledge components based on collected data from interviews, video recordings, herding diaries, and snow and temperature measurements to obtain, compare and amplify their understanding of climate change, adaptation mechanisms, and how to assess their risk in the field (Eira 2012, p. 129). In other words, a method was developed for monitoring climatic conditions by keeping diaries where snow concepts are combined with snow/temperature measurements together with snow and weather observations compiled and recorded in order to provide accurate assessment, communication, and interaction within the communities, based on the time and context the concept is used.

As important as predicting Arctic conditions accurately is the communication and application of this terminology and technicality in the field. Inuits and Sámis have the specialized terminology for snow and ice conditions consisting in specific concepts and vocabulary to describe the natural phenomena. This understanding based on the experience of the previous generations has enabled local Arctic communities to manage natural resources and survive for thousands of years in the harsh Arctic environment in a sustainable way. This cumulative experience from traditional way of life passed by cultural transmission consists in a local know-how to observe, predict, use and conserve natural resources, and provide constant adaptation for the sake of future generations. It is the fruit of an intergenerational daily attention to phenomena (such as dangerous ice) in the natural and social environments.

Adaptation depends on efficient communication that improves with a well-structured education and curriculum oriented to sustainable ways of development. The holistic dimension of the Arctic issues requires an equally holistic approach to education in a way that information can be disseminated in a variety of formats intelligible to different stakeholders.

2.6.1 Education in multicultural societies – Arctic indigenous groups in Norway, Canada, and Alaska

The Indigenous peoples of the Arctic or, the First Peoples, are very diverse encompassing distinct sub-groups and communities and accounting for approximately 10 % of the four million people living across the territories of eight Arctic nations – Norway, Sweden, Finland, the Kingdom of Denmark, Iceland, Canada, Russia, and the United States (AHDR-HH 2014).

According to Fondahl et al. (2015, p. 8) in 'Indigenous Peoples in the New Arctic':

> Of the estimated approximately four million people who inhabit the Arctic, approximately 10% are Indigenous. This proportion varies greatly across the Arctic. For instance, Inuit comprise about 85% of the population of Nunavut, Canada, and the great majority of Greenlanders are Indigenous as well, while in other areas, such as the Khanty-Mansi Autonomous Okrug of Siberia, Indigenous peoples make up less than 2% of the population.
>
> Alaska is home to numerous Indigenous peoples other than the Inuit, including the Aleut, Yup'ik, Tlingit and several Dene-language peoples (Ahtna, Deg Hit'an [Ingalik], Dena'ina [Tanaina], Doogh Hit'an [Holikachuk], Dichinanek Hwt'ana [Kolchan], Eyak, Gwich'in [Kutchin], Hän, Koyukon, Tanana). The Canadian North's Indigenous peoples include, as well as the Inuit, Cree peoples, several Denelanguage peoples (Deh cho [Slavey], Denesuline [Chipewayn], Dunneza [Beaver], Gwich'in, Hän, Kaska, Sahtu [Hare], Tagish, Tlicho [Dogrib], Tutchone,) and the Innu. The Russian Federation is homeland to over 40 Northern Indigenous groups, including as well as the Sámi and 'Eskimosy' (Inuit/Yuit), the Nentsy, Khanty, Mansi, Selkup, Evenki, Even, Dolgan, Yukaghir, Chukchi, Aleut, Itelmen, and numerous others. These people vary widely in language and culture. Within some of the more widely distributed peoples, sub-groups have strong place-based identities.
>
> (Fondahl et al. 2015, p. 8)

According to the First Nations Information Governance Centre (FNIGC-First Nations Information Governance Centre 2012, p. 42) there are 633 First Nation communities involving 11 language families and over 60 language dialects specific to local communities across Canada. The survey also reported that more than half (56.3%) of First Nations youth (on reserve) can understand or speak a First Nations language, and 45.8% feel that learning a First Nations language is very important (FNIGC-First Nations Information Governance Centre 2012, pp. 41, 42). In Canada, the predominant indigenous peoples are the First Nations. The northern communities have lived in parts of the Yukon, Northwest Territories, Nunavut, Nunavik (Quebec) and Nunatsiavut (Labrador). Despite the recognition of indigenous peoples' contribution to Canadian

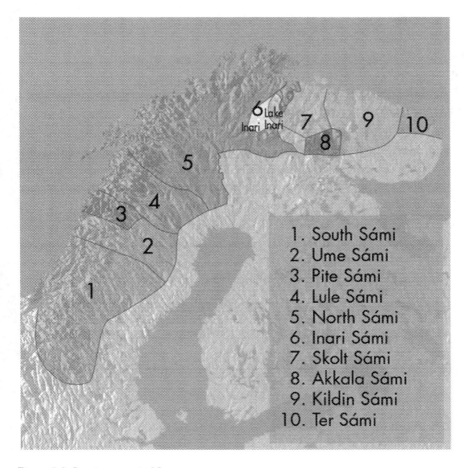

Figure 2.3 Sami groups in Norway.
Map source: SIIDA, ANARÂŠ at www.samimuseum.fi/anaras/english/kieli/kieli.html

society, there are still signs of the tension and mistrust from the days of the 'residential schools' and the 'truth and reconciliation' commission.[1] Sámis are the predominant demographic of indigenous population in Norway, but they are also present in other areas like Sweden, Finland, Russia, the United States, and Ukraine. In this study, the focus is on the indigenous populations from Norway, Canada, and Alaska (see Figures 2.3 and 2.4 and Table 2.1).

It is important to have an overview of the dimension of indigenous presence in the Arctic to understand the importance of rethinking Arctic HE, the social complexity, and the dimension of the sustainability (natural resources, energy) debate in the studied locations.

In Alaska, Alaskan natives (see Figure 2.4) include Inupiat, Yupik, Aleut, Eyak, Tlingit, Haida, Tsimshian, and northern Athabaskan distributed across 13 regional corporations according to the Alaska Native Claims Settlement Act (Kentch 2012).

Table 2.1 Northwest Territories First Nations and Yukon First Nations.

Northwest Territories' First Nations	Yukon First Nations
1. Acho Dene Koe	1. Little Salmon/Carmacks First Nation
2. Aklavik	2. Carcross/Tagish First Nations
3. Behdzi Ahda' First Nation	3. Champagne and Aishihik First Nations
4. Dechi Laot'i First Nations	4. Dease River First Nation
5. Deh Gah Gotie Dene Council	5. First Nation of Nacho Nyak Dun
6. Deline First Nation	6. Kluane First Nation
7. Deninu K'ue First Nation	7. Kwanlin Dun First Nation
8. Dog Rib Rae	8. Liard First Nation
9. Fort Good Hope	9. Ross River
10. Gameti First Nation	10. Selkirk First Nation
11. Gwichya Gwich'in	11. Ta'an Kwach'an
12. Inuvik Native	12. Taku River Tlingit
13. Jean Marie River First Nation	13. Teslin Tlingit Council
14. K'atlodeeche First Nation	14. Tr'ondëk Hwëch'in
15. Ka'a'gee Tu First Nation	15. Vuntut Gwitchin First Nation
16. Liidlii Kue First Nation	16. White River First Nation
17. Lutsel K'e Dene First Nation	
18. Nahanni Butte	
19. Pehdzeh Ki First Nation	
20. Salt River First Nation #195	
21. Sambaa K'e (Trout Lake) Dene	
22. Tetlit Gwich'in	
23. Tulita Dene	
24. West Point First Nation	
25. Wha Ti First Nation	
26. Yellowknives Dene First Nation	

Adapted from First peoples of Canada. Source: https://firstpeoplesofcanada.com/index.html

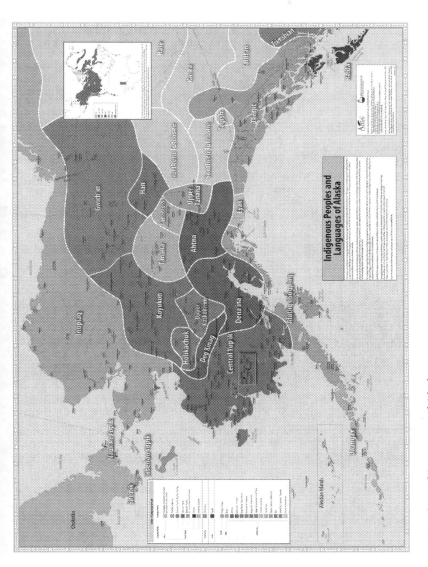

Figure 2.4 Indigenous Peoples and Languages of Alaska.

Source: Alaska Native Language Center and UAA's Institute of Social and Economic Research. Map courtesy of the Alaska Native Language Center at the University of Alaska Fairbanks (www.uaf.edu/anla and www.uaf.edu/anla/collections/map/)

The ethnical diversity in the Arctic areas under study demonstrates the linguistic and cultural complexity of Artic education. Education in multicultural societies is a right established by the UN General Assembly in 2007 through 'The United Nations Declaration on Indigenous Peoples' setting the rights of indigenous peoples to language, culture, identity, health, education, employment, and other critical issues (Wiessner 2007, p. 4). According to this declaration, indigenous peoples have also the right:

a) 'to freely determine their political status and freely pursue their economic, social and cultural development' (Article 3 – right to self-determination) (Wiessner 2007, p. 4);
b) 'to autonomy or self-government in matters relating to their internal and local affairs,' (Article 4 – self-government) (Wiessner 2007, p. 4);
c) 'to maintain and strengthen their distinct political, legal, economic, social and cultural institutions' (Article 5 – distinct institutions) (Wiessner 2007, p. 4).

Despite the United Nations Declaration on Indigenous Peoples' there is a tension between indigenous and non-indigenous styles of education because indigenous cultures have undergone substantial transformation due to the influence of state education and policies, language repression, the influence of exogenous religions, sedentarization, and forces of globalization (Fondahl et al. 2015). This tension is expressed by the existence of competing realities.

These different realities determine distinct sets of knowledge. Keskitalo (1998, p. 192) is explicit when arguing that science is a cultural activity. Western science is developed according to Western culture, Arctic knowledge is developed according to Arctic traditional way of life.

> The Sami have had access to elementary and higher education and to material welfare in general to an extent which indigenous peoples in the greater part of the world can hardly dream of attaining.
> (Keskitalo 1998, p. 192)

In the 20th century, a new understanding of social rights emerged according to which civil rights and the right to political participation are seen to be little worth for the citizen without a certain amount of economic security, general welfare, and education. In the 21st century, economic security continues to be central; however, the world offers new challenges that certainly will not be resolved by an outdated mechanistic, reductionist, and separatist worldview (Boulton et al. 2015). For the last two centuries, the world has been explained according to a mechanized way of thinking based on the industrial revolution and the trend to mechanize and polarize the own process of thinking. This mechanical worldview 'continues to maintain its attraction in Western societies as it provides a sense of order, purpose and control', but it does not seem appropriate for an Arctic reality of variation, change, adaptability, and creativity (Mayr 1991, p. 97).

Darwin's view of variation and selection leads to what he called 'path-dependent' emergence of new forms. It means that there is no sense of perfection, but what emerges builds on what comes before. This idea explains the way some Arctic communities understand knowledge and how they build up their intergenerational knowledge systems.

The explicit interest in the field of education by Sámi scholars is due to the growing recognition that, in order for this group to use their civil and political rights on an equal level with non-Sámi Norwegians, their access to elementary and HE is crucial. The field of education, however, is a good example of how different treatment sometimes is necessary to create equal opportunities. The indigenous education system (First Nations, Sámi, Dene, Inuit, etc.) is based on language and cultural activities. The right to receive HE in the Sámi language, and in accordance with Sámi culture and values, is not only a matter of equality but, agreeing with authors such as Honneth (1995, p. 129), Benhabib (2002, p. 82), a matter of constructing knowledge based in culture and cultural rights to protect the culture (Jakobsen 2011, p. 6). For many Arctic graduate students, it can be relatively problematic to find a validated methodology for indigenous research based on Western education system as it can impose patterns that exclude other forms of cultural knowledge that matters in a multicultural society. The culture-based knowledge construction introduces a new interdisciplinary methodology for indigenous scholars entitled 'Aboriginography', which is a decolonized, indigenous-centred research methodology that reflects the indigenous epistemology and pedagogy, philosophically centred on traditional teachings that are holistic in nature stating that learning occurs within the entire context of what is being studied (Longman 2014, p. 16).

Longman adds that:

> The model of Indigenous epistemology serves as a valuable template for researchers and teachers, in which the acquisition of new knowledge is achieved directly within the indigenous context, through engaged cultural immersion in indigenous communities, on and off reserve, through primary research data attained directly from Indigenous people, and through interviews or apprenticeships with indigenous experts. This framework is necessary in order to decolonize Eurocentric patterns of educational dominance and to widen definitions of knowledge. Today, cultural imperialism and domination over indigenous knowledge is no longer appropriate.
>
> (Longman 2014, p. 17)

This methodological model is focused on indigenous research produced by indigenous people according to a holistic philosophy based on the whole contextualization of cultural, social, and historical realities. It uses a qualitative research format derived from storytelling, cultural observation, diaries, photos, and interviews obtained from the natural and societal context. Knowledge in writing comes from indigenous sources, based on past and present personal experience and practice based on a dynamic balance where the contexts are changeable, not static as in mechanized Western patterns. In several parts of the Arctic, indigenous scholarship begins to be based on language, history, and culture, and the recent paradigmatic shift in educational institutions, indigenous curriculum, and faculty has had a greater presence in the HE. Longman (2014, p. 17), observes that the Canadians, apart from non-Sámi Norwegians, are now considering what can be learnt from the indigenous peoples and how this cultural gap can be bridged (Longman 2014, p. 17). In Canada, there are educational initiatives (Burton & Point 2006) that turned into frameworks for integrating indigenous and scientific knowledge, at academic level, to reduce vulnerability to environmental hazards, to increase collaboration among stakeholders and achieving risk reduction planning not only in relation to natural risks but also concerning human-induced risks like the energy industry exploration. Agius et al. (2004, p. 219) note that the aim of these frameworks is to promote indigenous active participation in decisions, enabling communities to reach consensus regarding ways to approach their vulnerability to risks. Some of the risks being examined are the ones associated to energy production and use according to Alberta Emerald Foundation (2008, pp. 78, 104). From the indigenous people's viewpoint, there are some specific expectations when enrolling to a HE course, as follows:

> (...) a higher educational system that respects them for who they are, that is relevant to their view of the world, that offers reciprocity in their relationships with others, and that helps them exercise responsibility over their own lives.
> (Kirkness & Barnhardt 1991, p. 10)

In general, indigenous students, regardless the ethnicity, are searching, in HE, for a space of learning that respects them for whom they are and for their values, perspectives, and cultural integrity. This is intrinsically linked to their innate international 'rights to land, territories and resources traditionally owned, used or acquired' (UNDRIP 2007, §26):

> Indigenous rights to land and resources is a fundamental issue in the Arctic, in that land and resources lie at the heart of both cultural and material well-being for Indigenous northerners. Many Indigenous peoples still heavily rely on the harvesting of biological resource – Arctic fauna and flora – for their subsistence needs. Strong ties to the land are repeatedly invoked as a key element of Arctic Indigenous cultural identity (AHDR 2004; ASI 2010). The International Labour Organization's Convention 169 (ILO 169), while also speaking to indigenous rights to land, has only been ratified by two Arctic states, Denmark and Norway. Thus, lack of recognition and unresolved questions of territorial rights

have been major hindrances to the economic and cultural well-being of these peoples. A rich source of hydrocarbons, minerals and fish, the Arctic is seeing increased interest in development, due to both to growing global demand for such resources and, in some areas, increased accessibility (or anticipation of such), related to climate change. Arctic tourism is also projected to grow over the next decades. These activities and their associated infrastructure compete with traditional land use, encroaching on the territories used for hunting, reindeer herding, and fishing. They contribute to habitat fragmentation, deleterious for Arctic fauna, and facilitate access by outsiders with concomitant increases in sports hunting, poaching, destructive ATV activity, etc.

(Fondahl et al. 2015, pp. 13, 14)

Indigenous peoples want to be part of the current debate; they want to have an active role in the process of decision-making in relation to exploration projects developed in their lands with a proper knowledge base to make decisions that are compatible with their livelihoods and traditions. This is an important justification to rethink and reframe HE in the Arctic because there is a set of references that need to match indigenous perspectives. This point is corroborated by Kirkness and Barnhardt (1991) when arguing that researchers and educators should re-examine education and literacy in the North in order to understand what is happening:

It is vital for planners and practitioners in the North to move beyond Southern and Western systems of hierarchy and status. In this way, the issue can be addressed with solutions developed from a Northern viewpoint. Freire (1998) describes the oppressed students' situation as a 'culture of silence' (Freire 1998), and Hooks (1989) states that 'moving from silence into speech is a revolutionary gesture (...) only as subjects can we speak.'

(Kirkness & Barnhardt 1991, p. 10)

In fact, what happens is a mismatch between the Western educational system and the indigenous perspective when trying to impose a transition on the institution's culture. It has several consequences for the education and learning process of indigenous groups. When indigenous students do not adapt to the levels of exigence, requirements, language, and Western perspective on issues, they are labelled as low-performers, low-retention and high-attrition groups:

The institutional response, when faced with these internally constructed and externally reinforced problems of inadequate achievement and retention, is usually to intensify the pressure on First Nations students to adapt and become integrated into the institution's social fabric, with the ultimate goal that they will be 'retained' until they graduate.

(Kirkness & Barnhardt 2001, p. 97)

Job opportunities alone are not sufficient as a motivation for First Nation students to go into HE. However, reasons for pursuing a university education may transcend the interest and well-being of the student as an individual to involve more collective welfare. According to Kirkness and Barnhardt (1991, p. 99):

> For First Nations communities and students, a university education can be seen as important for any of the following reasons:
>
> - realizing equality and sharing in the opportunities of the larger society
> - collective social and economic mobility
> - overcoming dependency and neo-colonialism
> - engaging in research to advance the knowledge of First Nations
> - providing the expertise and leadership needed by First Nations communities.
>   - demystifying mainstream culture and learning the politics and history of racial discrimination.
>
> (Kirkness & Barnhardt 1991, p. 99)

Multicultural education and research have experienced a considerable growth in several parts of the Arctic recently and a multicultural accommodating curriculum is an important measure to enhance a meaningful HE access to First Nations, Sámis, Inuit, Metis, and American indigenous peoples. An emerging group of indigenous scholars across the Arctic is contributing in this accommodation process with views of how knowledge is constructed and disseminated (Scollon 1981).

HE represents knowledge and empowerment to indigenous students. This notion of knowledge as power or powerful knowledge, which will be further explained in Section 3.2.4, was introduced by a British sociologist of education called Michael Young a decade ago. According to Young (2008, p. 14):

> Powerful knowledge refers to what the knowledge can do or what intellectual power it gives to those who have access to it. Powerful knowledge provides more reliable explanations and new ways of thinking about the world and (...) can provide learners with a language for engaging in political, moral, and other kinds of debates.
>
> (Young 2008, p. 14)

Apart from new ways of thinking about the world, Young (2013, p. 196) explains that:

> Powerful knowledge is powerful because it provides the best understanding of the natural and social worlds that we have and helps us go beyond our individual experiences.
>
> (Young 2013, p. 196)

> Knowledge is 'powerful' if it predicts, if it explains, if it enables you to envisage alternatives.
>
> (Young 2014, p. 74)

This book argues that these variations of the concept align with the idea of transition, alternative futures (Hicks 1991), global-local views, and the inclusiveness that young indigenous students currently search in HE to understand the complex contexts that they are part of (Arruda 2018a). Apart from the links between the natural and social worlds, the intellectual power and capabilities that the encounter of scientific knowledge and TK can provide through the practice of indigenous methodology represent an additional extraordinary component to innovate power knowledge in the Arctic geographical space.

2.7 Sustainable Development (SD) as a concept and Education for Sustainable Development (ESD)

As mentioned in Chapter 1 (Section 1.3) the way societies relate to energy influence their economic prosperity and levels of development. The limited multidisciplinary and intercultural understanding of undergraduates in relation to Sustainable Development and energy is observed in different parts of the Arctic and non-Arctic world. Projects in non-Arctic locations like the Regional Centres of Expertise on Education for Sustainable Development (RCE) from Hamburg, Germany (Leal Filho & Schwarz 2008, p. 499) or the Student and Teacher Value Assimilation (SATVA) in Bangalore, India (Sundaresan & Bavle 2008, p. 171) and the Systems Change Model applied in Australia (Davis & Ferreira 2008, p, 220); are examples of a comprehensive set of initiatives aimed at bringing Sustainable Development closer to learners' minds.

This need for improvement seems to occur due to a low curriculum adherence to UNESCO ESD, a framework developed to promote 'knowledge, skills, attitudes and values necessary to shape a sustainable future' (UNESCO 2014, p. 46). However, there seems to be additional factors contributing to this lack of clarity or engagement.

As introduced in Chapter 1 and expanded in Chapter 2, SD is defined by the United Nations Conference for Environment and Development (Rio 92) as 'that which meets present needs without compromising the ability of future generations to meet their own needs'. The term, Sustainable Development, was popularized in Our Common Future, a report published by the World Commission on Environment and Development (UNWCED 1987, p. 10). Also, known as the Brundtland Report, Our Common Future included the 'classic' definition of Sustainable Development in Principle 15 of the Rio Declaration on Environment and Development. Acceptance of the report by the United Nations General Assembly gave the term political salience along the

1990s and, in 1992, leaders set out the principles of SD at the United Nations Conference on Environment and Development.

How development has been understood along history presents significant changes and interfaces with other relevant aspects of modernization. In 1992, the concept encapsulated more the environmental aspect reflecting the trend of the 1990s related to the insufficiency of 'end-of-pipe' solutions to environmental problems. It opened the way to a more comprehensive understanding of development, in the 2000s, connected to human development and the achievement of decent standards of living, human capabilities, and the promotion of human well-being (UNDP 2001, p. 9).

Another fundamental component in this historical trajectory along the 2000s to understand the term development was the focus on the different living costs in several parts of the world, or the focus on poverty alleviation, and initiatives from international institutions like the World Bank shifting the development agenda to the Millennium Development Goals (MDGs) and, in 2015, to the Sustainable Development Goals (SDGs) (Black & White 2004, p. 10; Desai & Potter 2014, p. 25)

Several different interpretations have tried to uncover the meaning behind the concept, and it seemed to mean different things to different people (stakeholders) in different contexts (Baker et al. 1997), but one indissociable aspect is the connection between processes of industrialization and development. As an attempt to deepen the understanding of the concept and make it applicable in practice to businesses and industries, the so called 'Triple Bottom Line', or 'TBL' was a term coined by John Elkington, in 1994, as the outcome of an international expert's survey in corporate social responsibility (CSR) and SD by highlighting the importance to address, in a more integrated way, the social and economic dimensions of the environmental agenda initially proposed by 1987's Brundtland Report (UNWCED 1987, p. 10). In its genesis, it was understood as a possibility to improve the quality of human life inside the limits of the supporting capacity of the ecosystems.

TBL has been associated with the term 'sustainability' understood as a function of SD that also emerged from the business practice and the corporate social responsibility field, implying, as Elkington (1994, p. 92) argued, a 'win-win-win' business strategy. The concept can also be understood under the perspective of the 'Three P formulation', 'people, planet and profits', used for the first time in The Shell Report (1998, pp. 4, 53, 54) the main objective of which was to put the theory into practice by demonstrating that Shell Oil Company has values:

> At the heart of the emerging sustainable value creation concept is a recognition that for a company to prosper over the long-term it must continuously meet society's needs for goods and services without destroying natural and social capital.
>
> (The Shell Report 1998, p. 53)

The Arctic context 55

If sustainable development is to become a global reality rather than remain a seductive mirage, governments, communities, companies and individuals must work together to improve their 'triple bottom line' (economic, social and environmental) performance. To this end, we not only need new forms of accountability but also new forms of accounting.

(The Shell Report 1998, p. 54)

It was a significant landmark, as the term started to be intelligible and embedded into business strategy and leadership of an energy company, Shell. In the simplest terms, Elkington's TBL agenda focused not just on the economic value that corporations add, but also on the environmental and social value they add – or destroy (Henriques & Richardson 2004, pp. 23, 24). The drivers of the TBL agenda reflected a new trend aiming at addressing societal pressures on business (externalities), a shift in values that have inspired different business models and emerging roles of governments as per in Figure 2.5.

	Old Paradigm →	New Paradigm
1 Markets	Compliance →	Competition
2 Values	Hard →	Soft
3 Transparency	Closed →	Open
4 Life-cycle technology	Product →	Function
5 Partnerships	Subversion →	Symbiosis
6 Time	Wider →	Longer
7 Corporate governance	Exclusive →	Inclusive

Figure 2.5 Seven drivers of the Triple Bottom Line responsible for the seven sustainability revolutions.
Source: Henriques & Richardson 2004, p. 23

During the 2000s, the TBL original concept gained more clarity as, in 2009, Sustainable Cleveland 2019 Summit (2009) stated that:

Sustainability is a process and a way of thinking that helps us make better decisions about how to meet our needs. It helps us consider the long-term implications of our actions. It helps us appreciate the interdependence of a strong economy, a healthy environment, and vibrant society. It helps us develop a more resilient society that can adapt to a dynamic, changing world.

(Sustainable Cleveland 2019 Summit 2009, p. 9)

The concept that started as an attempt to apply theory into business practice evolved to become a true shift in the way of thinking, inspiring international initiatives across formal and informal learning contexts and at all educational levels. In education, it is more common to read of the 'three pillars of sustainability', but it is important to have in mind the historical background of the expression. The concept of sustainability focuses on achieving human well-being and quality of life, pursued through the maintenance, care, and equitable use of natural and cultural resources (Ryan 2011, p. 3). On the other hand, Hove (2004) argues that, unfortunately, Sustainable Development has become more of a catch phrase than a revolution of thought, justifying and reinforcing the paradigm that it initially sought to deconstruct. The great emphasis on economic growth undermines environmental reforms. In addition, the flawed and inherently contradictory ideology of SD may be improved by placing greater emphasis on social and economic equity (Fernando 2003, p. 31). Hove (2004, p. 53) also argues that:

> Sustainable Development represents an altogether vague, inherently contradictory approach to mediating the impasse to development. Three main critiques were made: 1) sustainable development is Western construct, perpetuating the ideological underpinnings of former approaches, 2) it focuses its efforts on the unsustainable expansion of economic growth, and 3) its broad nature creates dangerous opportunities for actors to reinterpret and mould the approach the way they see fit.
>
> (Hove 2004, p. 53)

Despite being a breakthrough in the 90s, not all scholars believe Sustainable Development is achievable, and the concept itself is contested. This is the reason, possibly, because ESD is not still the mainstream. ESD also referred as 'sustainability education' by Sterling (2001, 2004, 2008) was defined by UNESCO (2014, p. 46) as the education that 'allows every human being to acquire the knowledge, skills, attitudes and values necessary to shape a sustainable future'.

Ryan (2011, p. 3) also advocates that:

> ESD is a vision of education that seeks to balance human and economic well-being with cultural traditions and respect for the earth's natural resources. ESD applies transdisciplinary educational methods and approaches to develop an ethic for lifelong learning; fosters respect for human needs that are compatible with sustainable use of natural resources and the needs of the planet; and nurtures a sense of global solidarity.
>
> (Ryan 2011, p. 3; UNESCO 2005, p. 2)

The Arctic context 57

ESD, despite being an established concept endorsed by the United Nations Educational, Scientific and Cultural Organization (UNESCO) with the impetus of integrating ESD in different educational sectors and with relevant initiatives like the United Nation's World Decade on Education for Sustainable Development (2005–2014), is still not the educational mainstream. The main aim of ESD vision is to promote the balance between economic and human aspects of TBL, what has proved to be a difficult task so far. The ESD vision also advocates the compatibility of the human needs to the natural capacity of the planet and the fundamental sense of solidarity (compassion) that inspires the concept of citizenship approached in more depth in Chapter 3 of this book (see Section 3.4). The latter elements of respect to the natural capacity and solidarity are definitely not the mainstream in education and in civilization.

The policy dimension and historical background of ESD can be seen by the table of declarations, charters and programmes in HE from 1972 to 2014 extracted from Barth et al. (2016, p. 42).

Over the past 40 years of political activities in ESD, as per Table 2.2, the institutional framework evolved from an orientation and experimental phase (1970–1990) to a transition and development phase (1990–2000) to an expansionary phase (to 2014) revealing a trajectory from an environmental perspective to a sustainability perspective (Michelsen 2015). A key political activity was the United Nation's World Decade on Education for Sustainable Development (2005–2014), which is in progress since 2015 as Global Action Programme (GAP) in order to maintain support for the activities and actors of the previous initiatives by implementing ESD according to priority action areas like national-international ESD integration, whole-institution approaches, teacher education and training, youth involvement in educational processes, local-level ESD (Barth 2015; Leal Filho 2011; Michelsen 2011). However, these initiatives require further implementation and systematic application into practice in terms of educating the educators on a geographical basis and, consequently, enhancing the adherence of the curricula to ESD at all levels. International efforts to embrace the concept and principles could be seen along Rio+20 Conference, the Convention on Biological Diversity, the Sustainable Development Goals (SDGs), the UN Framework Convention on Climate Change (UNFCCC) and the 2030 Agenda for Sustainable Development. These are important initiatives materialized in international documents representing an important alignment to be followed by global education. As the focus of ESD resides on pedagogies and practices these are alignments that have inspired ESD initiatives as those illustrated in the beginning of Section 2.7 motivated by different drivers as professional performance, employability, personal and collective development, environmental management, etc.

Tilbury et al. (2005, p. 1) argues that ESD initiatives in HE tend to focus on single projects to address sustainability, as opposed to taking a more systematic view of learning and change across the institutions to

Table 2.2 The policy dimension and background of ESD.

Declaration/Charter/Programme	Initiator/Partner	Year	Keywords	Challenge	Approach
Declaration of the UN Conference on the Human Environment	United Nations	1972	Environmental Education in general	Environment	
Tbilisi Declaration	UNESCO/UNEP	1977	Environmental Education	Environment	Interdisciplinarity
Talloires Declaration	University Leaders for a Sustainable Future (ULSF)	1990	Higher Education Leadership for Sustainability	Environment Sustainability	Interdisciplinarity Transdiciplinarity
Halifax Declaration	Canadian Institutions IAU, UNU	1991	Ethical obligation Participation Leadership	Environment Sustainability	Whole-institution approach
Agenda 21	United Nations	1992	Chapter 36 Environmental Education	Environment Sustainability	
Kyoto Declaration	IAU	1993	Sustainability action plan Environmental Education	Environment Sustainability	Whole-institution approach
Swansea Declaration	Association of Commonwealth Universities	1993	Ethical obligation Environmental literacy and curriculum	Environment Sustainability	Whole-institution approach
COPERNICUS Charter	COPERNICUS and European Universities	1994	Embedding the environment and sustainability across higher education	Environment Sustainability	Interdisciplinary

World Declaration for Higher Education	UNESCO	1998	Responsibility Societal problems	Environmental Sustainability	Inter- and transdisciplinarity Whole-institution approach
Luneburg Declaration	GHESP	2001	Key role of universities Social change Curriculum reorientation	Sustainability	Whole-institution approach
Ubuntu Declaration	UNU, UNESCO, COPERNICUS GUESP, ULSF et al.	2002	Learning for sustainability Review of programs and curriculum MDGs	Sustainability	Whole-institution approach
Johannesburg Declaration/Johannesburg Plan	United Nations	2002	Education for Sustainable Development UN Decade ESD	Sustainability	
UNECE Strategy on ESD	UNECE	2005	ESD in general and in higher education	Sustainability	Interdisciplinarity Whole-institution approach
Graz Declaration	COPERNICUS, Oikos, UNESCO	2005	Embedding sustainability across higher education	Sustainability	Interdisciplinarity Whole-institution approach
Bologna Declaration/Bergen Communiqué	Ministers of Education in Europe	1998/2005	Sustainability Curriculum reform process	Sustainability	
Sapporo Sustainability Declaration	G8 University Network	2008	Need for global sustainability Reorientation of education and curriculum	Sustainability	

(*Continued*)

Table 2.2 (Cont.)

Declaration/Charter/Programme	Initiator/Partner	Year	Keywords	Challenge	Approach
Bonn Declaration	UNESCO	2009	ESD Higher education institutions	Sustainability	Multidisciplinarity Whole-institution approach
The Future We Want – Rio+20 Declaration	United Nations	2012	ESD Quality education	Sustainability	Interdisciplinarity
Rio +20 Treaty on Higher Education	COPERNICUS, MESA, GUNI, GUPES, IAU, ISCN, UNU, RCE et al.	2012	Responsibility of universities ESD in higher education Transformation research	Sustainability	
Aichi-Nagoya Declaration on Education for Sustainable Development	UNESCO	2014	ESD in general Responsibility Transformative knowledge	Sustainability	Whole-institution approach

Source: Barth et al. 2016, pp. 42–43

effectively addressing fundamental sustainability challenges (Mulà et al. 2017, p. 801).

Apart from projects and initiatives, another important aspect that is not much emphasized in HE is how to evaluate the adherence of courses, programs, and curricula to ESD. This is an important aspect of ESD's practical implementation considering the important role of HE in leading society towards a more sustainable future by providing tangible educational contributions to SD (Cortese 2003) when assessing courses alignment (adherence) to ESD principles.

Sterling (2001, 2005) was the precursor of 'sustainability education' by proposing types of sustain ability education in which transformative learning, and change of paradigms would challenge existing patterns and worldviews to construct new knowledge in order to have the capacity to deal with sustainability issues. Sterling (2001) proposed that education about sustainability tackles sustainability as a concept with no change of paradigm, while education for sustainability would focus on a reformulation of the curriculum and education as sustainability would involve life-long learning with a more transformative and participative character.

Following these premises, authors like Vare and Scott (2007) systematized levels of ESD in ESD 1 and ESD 2 as follows:

Table 2.3 Two sides of ESD.

ESD 1	ESD 2
• promoting/facilitating changes in what we do	• building capacity to think critically about [and beyond] what experts say and to test sustainable development ideas
• promoting (informed, skilled) behaviours and ways of thinking, where the need for this is clearly identified and agreed	• exploring the contradictions inherent in sustainable living
• learning for sustainable development	• learning as sustainable development

Source: Vare & Scott 2007, pp. 3, 4

The authors proposed a model in two levels – ESD1 and ESD 2 – providing characteristics of ESD in curricula being delivered through the relationship between educational outcomes (learning) and social change (behaviour change).

Also aligned to learning and change, Scott and Gough (2003, pp. 113, 116) proposed three types of ESD approaches, establishing a gradation, or as I call it in this research, possible 'levels of adherence' of ESD according to the following framework:

62 The Arctic context

Table 2.4 Three types of ESD.

Type 1 – learning about SD	Type 2 – learning for SD	Type 3 – learning as SD
• Assume that the problems humanity faces are essentially environmental	• Assume that our fundamental problems are social and/or political, and that these problems produce environmental symptoms	• Assume that what is (and can be) known in the present is not adequate
• Can be understood through science and resolved by appropriate environmental and/or social actions and technologies	• Such fundamental problems can be understood by means of anything from social-scientific analysis to an appeal to indigenous knowledge	• Desired 'end-states' cannot be specified
• It is assumed that learning leads to change once facts have been established and people are told what they are	• The solution in each case is to bring about social change	• This means that any learning must be open-ended
	• Learning is a tool to facilitate choice between alternative futures that can be specified on the basis of what is known in the present	• Essential if the uncertainties and complexities inherent in how we live now are to lead to reflective social learning about how we might live in the future

Source: Scott & Gough 2003, pp. 113, 116

Another relevant model for ESD learning is the UNESCO's five pillars of learning (or five pillars of ESD) that was proposed by Jacques Delors et al. (1996) in his book entitled *The Treasure Within* published by UNESCO. UNESCO's Five Pillars of Learning is an approach to education based on lifelong learning and it consists of:

a) **Learning to know** is about having a broad general knowledge and in-depth understanding of a small number of subjects;
b) **Learning to do** is about having a main occupation but being skilled to deal with different situations and to work in teams;
c) **Learning to live together** is about understanding other people and our interdependence;
d) **Learning to be** is about personal development to make better choices and become more responsible;

e) **Learning to transform oneself and society** is about individuals working separately and together to change the world. This means gaining the knowledge, values, and skills needed for transforming attitudes and lifestyles (Delors et al. 1996, p. 37).

The complexity and unpredictability of current global challenges require from communities a higher level of understanding and sustainability. The Arctic cross-cultural environments require from people more than 'learning to know' and 'learning to do', but 'learning how to live together' in the learning site, in society, in the region by minimizing economic, social, environmental and cultural tensions and by understanding the interdependence of living sustainably and responsibly in one unique planet. Figure 2.6 shows the pillars of learning, in other words, the essence of how learning should be. Learning is more than knowing, it is an expression of the being and the well-being; it is inclusive, collaborative, engaging, and transformative to the individual and to society through the transformation of attitudes and lifestyles.

In this sense, Haigh (2014, p. 50) and Mezirow (1997, p. 7) seem to have captured the essence of the issue when saying that 'we do not make transformative changes in the way we learn as long as what we learn fits comfortably in our existing frames of reference'. Transformative learning is the process by which these frames of reference are changed, ideally, towards something more inclusive and integrative (Mezirow 1997, p. 7). These frames of reference can expand or be transformed when different perspectives are incorporated to the learning experience.

When dealing with intercultural education (indigenous and non-indigenous institutions) the learning practice would require dealing with multiple perspectives. This is a position advocated by Beck and Cowan (1996, p. 34) in their so-called 'memes' or the power of the evolutionary development of groups beyond the individual to the collective under a spiral model depicting the 'emergence of human systems' and 'levels of increasing complexity' (Beck 2002, p. 9). In this sense, this book argues that transformative learning is crucial in intercultural education because:

- Learners must be open to change, not locked into a particular worldview;
- Learners must be aware of dissonance, unresolved problems in their current frame;
- Learners must feel to explore the possibility of resolving these problems through the construction of a new mental framework;
- Learners must have the opportunity space needed to explore alternatives and develop his/her own personal insights;
- Leaners must confront and overcome barriers – psychological, emotional, and socio-cultural – that lock him/her into their present mental framework;
- Leaners must be given support during time when they are building and consolidating any new understanding (Beck & Cowan 1996, p. 34).

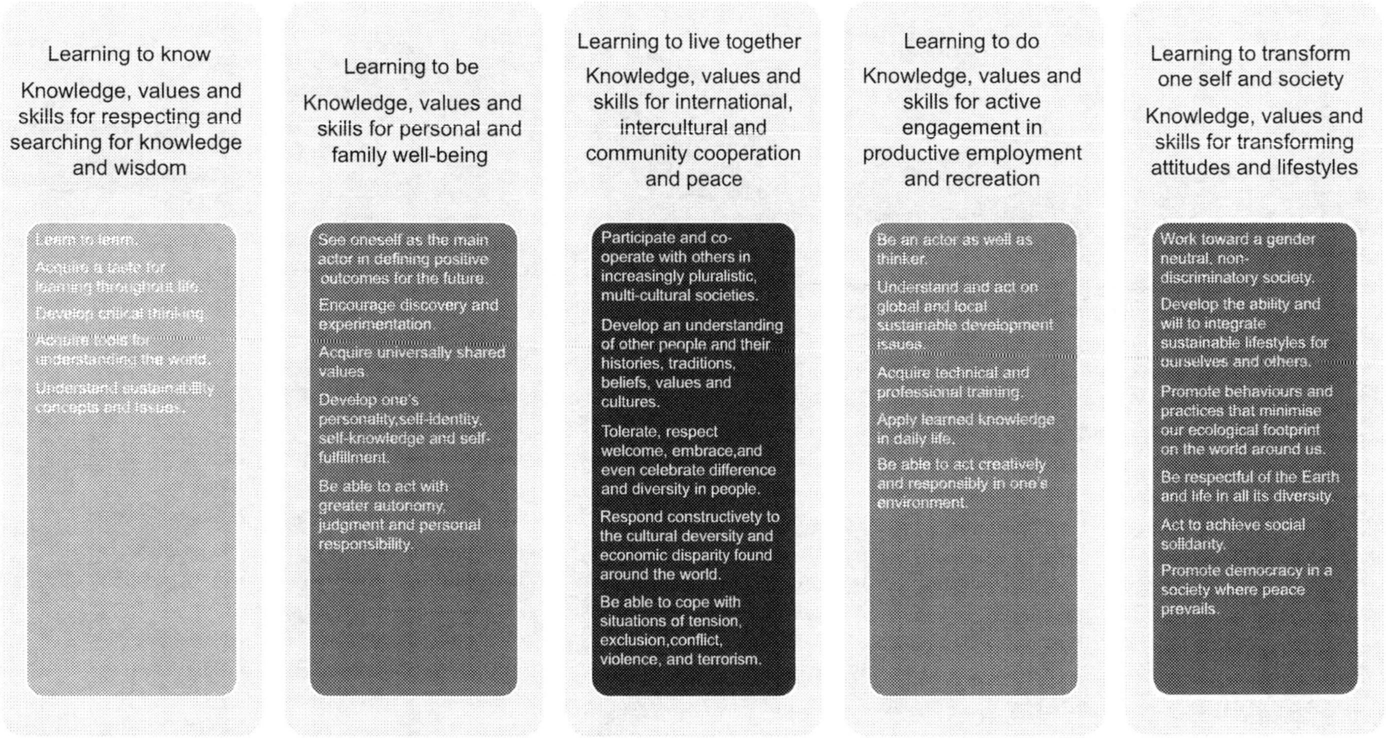

Learning to know	Learning to be	Learning to live together	Learning to do	Learning to transform one self and society
Knowledge, values and skills for respecting and searching for knowledge and wisdom	Knowledge, values and skills for personal and family well-being	Knowledge, values and skills for international, intercultural and community cooperation and peace	Knowledge, values and skills for active engagement in productive employment and recreation	Knowledge, values and skills for transforming attitudes and lifestyles
Learn to learn. Acquire a taste for learning throughout life. Develop critical thinking. Acquire tools for understanding the world. Understand sustainability concepts and issues.	See oneself as the main actor in defining positive outcomes for the future. Encourage discovery and experimentation. Acquire universally shared values. Develop one's personality, self-identity, self-knowledge and self-fulfilment. Be able to act with greater autonomy, judgment and personal responsibility.	Participate and co-operate with others in increasingly pluralistic, multi-cultural societies. Develop an understanding of other people and their histories, traditions, beliefs, values and cultures. Tolerate, respect, welcome, embrace, and even celebrate difference and diversity in people. Respond constructively to the cultural deversity and economic disparity found around the world. Be able to cope with situations of tension, exclusion, conflict, violence, and terrorism.	Be an actor as well as thinker. Understand and act on global and local sustainable development issues. Acquire technical and professional training. Apply learned knowledge in daily life. Be able to act creatively and responsibly in one's environment.	Work toward a gender neutral, non-discriminatory society. Develop the ability and will to integrate sustainable lifestyles for ourselves and others. Promote behaviours and practices that minimise our ecological footprint on the world around us. Be respectful of the Earth and life in all its diversity. Act to achieve social solidarity. Promote democracy in a society where peace prevails.

Figure 2.6 Five pillars of ESD.
Source: UNESCO 2005, p. 2

Haigh (2014, pp. 50, 51) states that transformative learning is an expansion of consciousness and the authentic and in-depth engagement with a continuous process of transformation implying the construction and reconstruction of the essence of the learner's subject of study or in other words a 'pedagogy of freedom' (Freire 1998, p. 33). This freedom allows a new mental framework, more expanded and coherent with concepts, values, beliefs, associations, and experience according to a construction based on more than one perspective or on more than one set of references. This provides a special eye to see 'the other' and to put yourself in the other's shoes.

The complexity around sustainability serves as an educational impulse for the improvement of learning processes but the ESD adherence and the metrics of these learning processes' outcomes require further development. The continuous search for sustainability presents an opportunity to develop learning activities and to explore debates over the issues at stake according to broader aims and approaches, experiential learning, critical thinking, reflection, and critical pedagogies. The critical point continues to be the practice, as Ryan (2011, pp. 3, 5) observes:

> HE is ideally positioned to make a critical contribution to sustainable development through its core academic functions of research and teaching (HEFCE 2009). However, while sustainability research has accelerated in recent years, curriculum development to date has been limited in scope and impact, due to the complexities of sustainability when applied within the existing academic structures and processes of HE. The field of education for sustainable development (ESD) has the strategic aim of reorienting entire educational systems, which in HE means the challenging goal of achieving large-scale shifts of curriculum priorities, policy, and practice.
>
> (Ryan 2011, p. 5)

This is the point when geographical knowledge defines the strategy, the educational system, and the curriculum, because this book argues that the 'cultural component' makes a huge difference in the SD equation despite it is still not considered as a pillar of the Triple Bottom Line (TBL). Cultural concerns are missing from the Sustainable Development concept and from ESD. The cultural pillar means how to see the 'other' in his/her integrity, to respect and interact with others peacefully and considering different perspectives. It is not only about cross-cultural competences but also about the right to develop one's own culture and opinions. This is an important contribution of this research as the complexity of the geographical Arctic landscape, in its several components – economic, social, environmental, and cultural – defines specific approaches, pedagogies, educational experiences, and curriculum, as seen in Section 2.6.1, in relation to the local reality of Sustainable Development.

66 *The Arctic context*

Sustainable Development Literacy (eco-literacy) builds upon a progression of environmental and ecological literacies (UNESCO 2015, p. 8). The basic premise of Sustainable Development Literacy (SDL) is that human and natural systems are dynamically interdependent and cannot be considered in isolation in order to resolve critical issues through the application of process-based tools capable of managing unexpected change (Hutchins 1968, p. 1). As the geographical and cultural aspects are also relevant to define models of development, they seem to be equally important to define levels of Sustainable Development Literacy (Dale & Newman 2005, p. 352).

The critical element of change in the Arctic is energy production/use in sensitive areas impacting human and natural systems as well as the (in)capacity of understanding these geographical changes through HE. Terminology around sustainability is complex and culturally contested, with definitions varying according to context and perspective. Petrov et al. (2016, p. 165) argues that:

> Understanding the sustainability of Arctic social–ecological systems (SES), their ability to respond to external and internal pressures, and their adaptive capacities has become a key task for both researchers and Arctic stakeholders.
> (Petrov et al. 2016, p. 165)

Petrov (2014, p. 14) also proposes a different definition of Sustainable Development for the Arctic context as the 'development that improves health, human development and well-being of Arctic communities and people while conserving ecosystem structures, functions and resources' (Petrov 2014, pp. 14, 15). This seems to coincide with the idea that Education and Sustainable Development have an interdependent relationship as human development is promoted through the adoption of appropriate standards oriented to values that play a fundamental role in the process of deliberating on well-being (Larsen & Fondahl 2015, p. 500; Poppel 2015, p. 67; Sen 1993, p. 31).

If the main concern has been to expand the understanding of sustainability making it more applicable and connected to reality, Petrov (2014) and Poppel's (2015) perspectives seem to realize this aim as their concepts consider the context-specific factors (geo-multicultural factors) of the Arctic reality. This idea justifies the significance of developing geo-capabilities through HE curricula, a specific branch of the capabilities approach to human development and welfare economics developed by Amartya Sen and further developed by Martha Nussbaum (Nussbaum 2011, p. 33; Sen 2004, p. 16) which is described in Chapter 3 (see Section 3.3).

Finally, for decades the concept of SD with its three pillars seemed not to be clear enough for most people as an on-going process dependent on the geographical and cultural characteristics of models of development.

'Geographical' and 'cultural' are the two new pillars or variables proposed by this study (presented in the results chapters) in order to expand the understanding of the SD concept in a complex Arctic context. This context is being changed by human-induced activities, among them energy exploration as the genesis of all the human development debate.

2.8 Energy literacy and Arctic sustainable energy literacy

Energy education or energy literacy is a central component of Sustainable Development Literacy (SDL) and ESD. Apart from being the Goal 7 (Affordable and Clean Energy) of the UN General Assembly (2015, p. 14) energy is in the genesis of the most complex and dynamic sustainability challenges of our times as explained in Chapter 1. Cotton et al. (2015) describes accurately the challenge of confronting this topic when stating the relatively little attention given to energy literacy in the research literature, in general, despite the importance of climate change as an issue of international concern and despite the geographical impacts of energy production and use. Authors working on this subject, mainly in HE, are pioneers of a relatively new field with a narrow curriculum coverage about simplistic aspects of energy (Van Treuren & Gravagne 2008) and insufficient co-relation to human geography and human development, despite the US National Education Standards referring to energy literacy as a means of being literate in both natural and social sciences.

There are some studies covering this topic and conceptualizing energy literacy like DeWaters and Powers (2011, p. 1700) conceptualizing energy literacy in relation to secondary school courses, as comprising three domains; cognitive (knowledge), affective (attitudes and values), and behavioural and Cotton et al. (2015, p. 456) and Dwyer (2011a, p. 1) defining energy literacy as a 'construct that combines conceptual fluency with the economic and social components of energy use, along with the belief that an increase in energy literacy will result in more sustainable energy practices'.

Pilot projects related to energy literacy started in US secondary schools (NTET 2013) and evolved to a current framework applied to all ages (inclusive of HE) developed by the Energy Literacy Framework from the US Department of Energy. Energy literacy is an understanding of the nature and role of energy in the universe and in our lives. It is also the ability to apply this understanding to answer questions and solve problems (US Department of Energy 2017, p. 1).

This type of literacy matters, more than ever, because it influences decision-making, energy costs reduction, more informed decisions, national security, economic development, the reduction of negative environmental impacts, but mainly it promotes a sustainable energy model by disseminating understanding about renewable energy, energy efficiency and energy conservation exerting a relevant impact on quality of life of individuals and societies through their energy choices in different bio-regions. The problem is that the general limited knowledge of energy-related issues represents a barrier to participating in energy-related discussions and decision-making. In practical terms, low levels of energy literacy may reduce the ability of societies operating the transition to a low carbon economy and a low carbon energy system.

Energy literacy dynamics of teaching and learning present some additional challenges apart from the curriculum deficiency in terms of content, design, and activities to engage learners. It takes time for learners to reformulate their (economic and cultural) pre-conceived ideas about energy and educators would need to pay attention to a progressive learning experience in which energy is over time approached at different levels from simple to more sophisticated knowledge levels. Moreover, a crucial aspect of energy learning is the natural attachment it has to practice (Eidelman 2010) because natural and social sciences are bodies of knowledge interdependent of sets of practices (US Department of Energy 2017a, p. 13) in a continuous process of refinement. The US Department of Energy emphasizes, in its educational materials, the importance of connecting theory to practice and connecting the learners' personal experiences to energy topics to maximize the learning process and lifelong learning. Another relevant issue is that there are no defined methods to measure the efficacy of the learning process or what is argued in this book as the levels of curriculum adherence to energy literacy.

More useful for the purpose of this book and the frameworks explained in Chapter 3 are the components that can be found in the definition of an energy literate person that provides a more comprehensive concept, as follows:

> (…) one who has a sound conceptual knowledge base as well as a thorough understanding of how energy is used in everyday life, understands the impact that energy production and consumption have on all spheres of our environment and society, is sympathetic to the need for energy conservation and the need to develop alternatives to fossil fuel-based energy resources, is cognizant of the impact that personal energy-related decisions and actions have on the global community, and – most importantly – strives to make choices and exhibit behaviours that reflect these attitudes with respect to energy resource development and energy consumption.
> (DeWaters & Powers 2011, p. 1700)

There is a growing discussion about energy literacy as a component of SDL involving competencies (Stibbe 2009) and behaviours but it is important to highlight that the focus of energy literacy is not only to understand the impacts of extractive industries or to measure the energy consumption at home or on the university campus. It has a much bigger dimension and requires a much more consistent approach in education.

Transposing this discussion to an Arctic scale, energy literacy becomes even more complex to design and deliver because sustainable energy literacy (SEL) depends primarily on the transition to renewable energy in the Arctic. It depends on a sustainable energy model that is in formation and in progression, in other words, it is a process under development. In 'Renewable Energy for the Arctic: New Perspectives', the most important renewable energy projects and sources currently developed in the Arctic were mapped (Arruda 2018a). This pioneering book aimed at discussing the development of a sustainable energy system (model) for the Arctic, by analysing the challenges, opportunities and trends of how to shape and implement a low carbon Arctic energy system through modalities of renewable energy and forms of integrating energy in the Arctic, coordinated by co-management systems and intercultural interactions.

Other educational initiatives related to sustainable energy systems in the Arctic have been recently developed in Norway by the Arctic Centre for Sustainable Energy at UiT (The Arctic University of Norway) in 2017. Pilot educational projects have also been developed in Canada, in the Northwest Territories (NWT), consisting in a new energy strategy started in 2017 to develop and implement sustainable energy plans to reduce reliance on diesel in Indigenous and Northern Communities (United States delegation of the Arctic Council 2016). Both initiatives have adopted a more proactive strategy of education and 'best practices' for adaptation based on resilience priorities, challenges, and successes across Arctic states. In Alaska, energy education is also in progress with courses emphasizing how to decrease fossil fuel consumption in different environments as well as the application of renewable energy resources, energy efficiency strategies and a range of topics associated to technology, smart grids, integration, human development, and a low carbon Artic energy system vision (ACEP 2017).

SEL is not yet an Arctic-wide trend but the development and implementation of SEL depends on the bio-region characteristics (economic, social, environmental, cultural), on the local/regional/national energy models in place, on the initiatives to use the local renewable energy potential (geographical component) and on the co-evolving technical and social dynamics (Arruda & Arruda 2018, p. 102).

Note

1 Residential schools and the truth and reconciliation commission are part of the Canadian indigenous peoples' history. From 1831 to 1998, indigenous children were forcibly placed in residential schools operated across Canada. The schools were located near reserves and run by churches. The truth and reconciliation commission had the particular role of restoring the relationship between indigenous peoples and colonizers in a post-colonial society (Castellano et al. 2008, p. 183). The legacy of residential schools has weighed heavily on the lives and wellbeing of First Nations, Inuit, and Métis individuals and communities for generations.

References

AACHC (Alaska Arctic Council Host Committee) 2016, *Environmentally responsible resource use and development in the U.S. Arctic*, Alaska Arctic Council Host Committee, October 2016, p. 14. Alaska Arctic Council Host Committee.

Aberfeldy, SB & Smith, JD 2004, 'Building sustainable communities of practice', in PM Hildrethand & C Kimble (eds.), *Knowledge networks: innovation through communities of practice*, IGI Publishing, Pennsylvania, PA, pp. 150–164.

ACEP (Alaska Center for Energy and Power) 2017, 'Alaska Hydrokinetic Energy Research Center', viewed, http://acep.uaf.edu/programs/alaska-hydrokinetic-energy-research-center.aspx.

Adger, WN & Kelly, PM 1999, 'Social vulnerability to climate change and the architecture of entitlements', *Mitigation Adaptation Strategies Global Change*, vol. 4, pp. 253–256.

Agius, P, Davies, J, Howitt, R, Jarvis, S & Williams, R 2004, 'Comprehensive native title negotiations in South Australia', in M Langton, M Teehan, L Palmer & K Shain (eds.), *Honour among nations? Treaties and agreements with indigenous people*, Melbourne University Press, Melbourne, pp. 203–219.

Agrawal, A 1995, 'Dismantling the divide between indigenous and scientific knowledge', *Development and Change*, vol. 26, pp. 413–439.

AHDR 2004, *Arctic human development report*. Stefansson Arctic Institute, Akureyri.

AHDR-HH 2014, *Arctic human development report II*, Norden, Copenhagen.

Alberta Emerald Foundation 2008, *Background on youth environmental engagement in Alberta: how to engage youth in environmental action, and a snapshot of current activities*, Alberta Emerald Foundation, Alberta, Canada, pp. 78, 104.

AMAP 2007, *Arctic oil and gas 2007*, Arctic Monitoring and Assessment Programme (AMAP), Oslo, Norway, p. 10.

Andrachuk, M & Pearce, T 2010, 'Vulnerability and adaptation in two communities in the inuvialuit settlement region', in GK Hovelsrud & B Smit (eds.), *Community adaptation and vulnerability in Arctic regions*, Springer Science, London, pp. 64, 65, 68, 72.

Arruda, GM 2015, 'Arctic governance regime: the last frontier for hydrocarbons exploitation', *International Journal of Law and Management*, vol. 57, no. 5, pp. 498–521.

Arruda, GM 2018a, *Renewable energy for the Arctic: new perspectives*, Routledge, Abingdon.

Arruda, GM 2018b, 'Artic resource development. A sustainable prosperity project of co-management', in GM Arruda (eds.), *Renewable energy for the Arctic: new perspectives*, Routledge, Abingdon, p. 112.

Arruda, GM & Arruda, FM 2018, 'Towards sustainable energy systems through smart grids in the Arctic', in GM Arruda (ed.), *Renewable energy for the Arctic: new perspectives*, Routledge, Oxford, p. 102.

Arruda, GM & Krutkowski, S 2016, 'Arctic governance, indigenous knowledge, science and technology in times of climate change: self-realization, recognition, representativeness', *Journal of Enterprising Communities: People and Places in the Global Economy*, vol. 11, no. 4, pp. 514–528.

Arruda, GM & Krutkowski, S 2017, 'Social impacts of climate change and resource development in the Arctic: implications for Arctic governance', *Journal of*

Enterprising Communities: People and Places in the Global Economy, vol. 11, no. 2, pp. 277–288.

ASI 2010, *Arctic social indicators report*. Norden, Copenhagen.

Baker, S, Kansis, M, Richardson, D & Young, S 1997, *The politics of sustainable development*, Routledge, London.

Barth, M 2015, *Implementing sustainability in higher education. Learning in an age of transformation*, Routledge, London.

Barth, M, Michelsen, G, Rieckmann, M & Thomas, I 2016, *Routledge handbook of higher education for sustainable development*, Routledge, London, pp. 42, 43.

Beck, D 2002, Interview with Jessica Roemischer: 'the never-ending upward quest', *What Is Enlightenment*, vol. 22, pp. 4–22, 9.

Beck, D & Cowan, C 1996, *Spiral dynamics: mastering values, leadership and change*, Blackwell Publishers, Inc., Malden, MA, p. 34.

Benhabib, S 2002, *The claims of culture equality and diversity in the global era*, Princeton University Press, Princeton, NJ, p. 82.

Berkes, F & Jolly, D 2002, 'Adapting to climate change: social-ecological resilience in a Canadian Western Arctic community', *Conservation Ecology*, vol. 5, no. 2, p. 18.

Berkman, PA & Vylegzhanin, AN 2010, *Environmental security in the Arctic Ocean*, Springer, Cambridge, p. 182.

Berkman, PA & Young, OR 2009, 'Governance and environmental change in the Arctic', *Science*, vol. 324, no. 5925, pp. 339–340.

Bitz, C, Blockley, E, Kauler, F, Petty, A, Massonet, F, Arruda, GM, Sun, N & Druckenmiller, M 2016, *Post-season report publishing*, Washington, DC, viewed January 2017, www.arcus.org/sipn/sea-ice-outlook/2016/post-season.

Blaauw, RJ 2013, 'Oil and gas development and opportunities in the Arctic Ocean', in PA Berkman & AN Vylegzhanin (eds.), *Environmental security in the Arctic Ocean*, NATO Science for Peace and Security Series C: Environmental Security, Springer, Milton Keynes, UK, p. 180.

Bond, TC, et al. 2013, 'Bounding the role of black carbon in the climate system: a scientific assessment', *Journal of Geophysical Research: Atmospheres*, vol. 118, pp. 5380–5552.

Boulton, J, Bowman, C & Allen, P 2015, *Embracing complexity: strategic perspectives for an age of turbulence*, Oxford University Press, Oxford, UK, pp. 1, 10, 11, 20, 29, 35, 63, 66, 117.

Bridge, G, Bouzarovski, S, Bradshaw, M & Eyre, N 2013, 'Geographies of energy transition: space, place and the low-carbon economy', *Energy Policy*, vol. 53, pp. 331–340.

Briggs, J 2005, 'The use of indigenous knowledge in development: problems and challenges', *Progress Is Development Studies*, vol. 5, no. 2, pp. 99–114, p. 10, viewed 2 October 2016, http://eprints.gla.ac.uk/1094/1/JBriggs_eprint1094.pdf.

Burton, W & Point, G 2006, 'Histories of aboriginal education in Canada', in T Fenwick, T Nesbit & B Spencer (eds.), *Contexts of adult education: Canadian perspectives*, Thompson Educational Publishing, Toronto, pp. 36–48.

Castellano, MB, Archibald, L & DeGagné, M 2008, *From truth to reconciliation. The legacy of residential schools*, The Aboriginal Healing Foundation, Ottawa, Ontario, p. 183.

CCHRC (Cold Climate Housing Research Center) 2016, 'Annual report', viewed 27 November 2017, www.cchrc.org/sites/default/files/docs/2016AnnualReport.pdf.

Constitution Act 1867, 'Justice laws website. Department of Justice Canada', viewed 21 March 2019, https://laws-lois.justice.gc.ca/eng/Const/page-1.html.

Cortese, A 2003, 'The critical role of higher education in creating a sustainable future', *Planning for Higher Education*, vol. 31, no. 3, pp. 15–22.

Cotton, D, Miller, W, Winter, J, Bailey, I & Sterling, S 2015, 'Developing students' energy literacy in higher education', *International Journal of Sustainability in Higher Education*, vol. 16, no. 4, pp. 456–473.

Dale, A & Newman, L 2005, 'Sustainable development, education and literacy', *International Journal of Sustainability in Higher Education*, vol. 6, no. 4, pp. 351–362.

Davis, J & Ferreira, J 2008, 'Creating change for sustainability in universities in Australia, one system at a time', in W Leal Filho, L Brandli, P Castro & J Newman (eds.), *Handbook of theory and practice of sustainable development in higher education*, vol. 1, Springer, Hamburg, p. 220.

Delors, J, et al. 1996, *Learning: the treasure within: report to UNESCO of the international commission on education for the twenty-first century*, UNESCO, Paris, p. 37.

Desai, V & Potter, RB 2014, *The companion to development studies*, 3rd edn, Routledge, London, p. 25.

DeWaters, JE & Powers, SE 2011, 'Energy literacy of secondary students in New York State (USA): a measure of knowledge, affect, and behavior', *Energy Policy*, vol. 39, pp. 1699–1710.

Duhaime, G & Caron, A 2017, 'The economy of the circumpolar Arctic', in S Glomsrød, G Duhaime & I Aslaksen (eds.), *The economy of the north 2015*, Statistics Norway, Oslo, Norway, pp. 17–25.

Dwyer, C 2011a, 'Developing an energy literacy curriculum in support of sustainability', p. 1, viewed 5 January 2017, http://ssrn.com.oxfordbrookes.idm.oclc.org/abstract=1801463.

EIA/DOE (Energy Information Agency, US Department of Energy) 2017, 'State energy data system', viewed 20 December 2017, www.eia.gov/state/seds/data.php.

Eidelman, LA 2010, *Energizing outdoor environmental education: criteria and curriculum for outdoor energy education*, Book submitted for the degree of Master of Arts in Education. Hamline University, Saint Paul, MN (April 2010).

Eira, IM 2012, 'The silent language of snow. Sami traditional knowledge of snow in a time of climate change', Dissertation for the degree of Philosophiae Doctor, Faculty of Humanities, University of Tromso, January 2012, p. 129.

Elkington, J 1994, 'Towards the sustainable corporation: win-win-win business strategies for sustainable development', *California Management Review*, vol. 36, no. 2, pp. 90–100.

Elkington, J 2004, 'Enter the triple bottom line', in A Henriques & J Richardson, (eds.), *The triple bottom line: does it all add up? Assessing the sustainability of business and CSR paperback*, Earthscan, London, pp. 23, 24.

Fernando, JL 2003, 'The power of unsustainable development: what is to be done?', *The Annals of the American Academy*, vol. 590, pp. 6–31.

FNIGC (First Nations Information Governance Centre) 2012, *First nations regional health survey (RHS) 2008/10: national report on adults, youth and children living in first nations communities*, FNIGC, Ottawa, pp. 41, 42.

Fondahl, G, Filippova, V & Mack, L 2015, 'Indigenous people in the Arctic', in B Evengård, JN Larsen & Ø Paasche (eds.), *The new Arctic*, Springer International Publishing, Switzerland, pp. 8, 13, 14.

Forbes, BC 1995, 'Effects of surface disturbance on the movement of native and exotic plants under a changing climate', in TV Callaghan, U Molau, MJ Tyson, JI Holten, WC Oechel, T Gilmanov, B Maxwell & B Sveinbjornsson (eds.), *Proceedings of the International Conference 'Global change and arctic terrestrial ecosystems'*, 21–26 August 1993, Oppdal, Norway, Ecosystems Research Report, European Commission, p. 372.

Freire, P 1998, *Pedagogy of freedom: ethics, democracy and civic courage*, Rowan & Littlefield, Lanham, MD, p. 33.

Government of Nunavut 2007, 'Inuit Qaujimajatuqangit Education Framework'. Nunavut Department of Education, Curriculum and School Services Division, Nunavut, p. 23.

Haigh, M 2014, 'Gaia: "thinking like a planet" as transformative learning', *Journal of Geography in Higher Education*, vol. 38, no. 1, pp. 49–68.

Henriques, A & Richardson, J 2004, *The triple bottom line, does it all add up?* Earthscan, London, pp. 23, 24.

Hicks, D 1991, 'Preparing for the millennium: reflections on the need for futures education', *Futures*, vol. 23, no. 6, pp. 623–636.

Holland-Bartels, L & Pierce, B 2011, 'An evaluation of the science needs to inform decisions on outer continental shelf energy development in the Chukchi and Beaufort Seas, Alaska', U.S. Geological Survey Fact Sheet 2011-3048, p. 4, viewed 20 December 2016, http://pubs.usgs.gov/fs/2011/3048/.

Honneth, A 1995, *The struggle for recognition: the moral grammar of social conflicts*, Polity Press, Cambridge, p. 129.

Hove, H 2004, 'Critiquing sustainable development: a meaningful way of mediating the development impasse?', *Undercurrent*, vol. I, no. 1, pp. 48–54, 53.

Hovelsrud, G & Smit, B 2010, *Community adaptation and vulnerability in Arctic regions*, Springer, Dordrecht, New York, pp. 3, 69, 71, 75.

Hutchins, R 1968, *The Learning Society*, University of Chicago Press, Chicago, IL, p. 1.

IEA 2013, 'Resources to reserves. Oil, gas and coal technologies for the energy markets of the future', p. 18, viewed 10 January 2016, www.iea.org/publications/freepublications/publication/Resources2013.pdf.

Jakobsen, J 2011, 'Education, recognition and the Sámi people of Norway', in H Niedrig & C Ydesen (eds.), *Writing postcolonial histories of inter-cultural education*, Peter Lang, Frankfurt am Main, p. 6.

Kentch, G 2012, 'A corporate culture? The environmental justice challenges of the Alaska native claims settlement act', *Mississippi Law Journal*, vol. 81, no. 4, pp. 813–837.

Keskitalo, JH 1998, 'The saami experience: changing structures for learning', in L King (ed.), *Reflecting visions. New perspectives on adult education for Indigenous peoples*, UNESCO Institute for Education. The University of Waikato, Hamburg, Germany, pp. 187, 192.

Kirkness, V.J. and Barnhardt, R. 1991, 'First Nations and Higher Education: The Four R's – Respect, Relevance, Reciprocity, Responsibility', *Journal of American Indian Education*, vol. 30, no. 3, pp. 1–15.

Kirkness, V.J. and Barnhardt, R. 2001, 'First Nations and Higher Education: The Four R's – Respect, Relevance, Reciprocity, Responsibility'. In Hayoe, R. and Pan, J. 2016, Knowledge Across Cultures: A Contribution to Dialogue Among Civilizations. *The Journal of College and University Student Housing*, vol. 42, no. 2, Hong Kong, Comparative Education Research Centre, The University of Hong Kong, pp. 97, 99.

Larsen, JN & Fondahl, G 2015, *Arctic human development report: regional processes and global linkages*, Nordisk Ministerråd, Copenhagen, p. 500.

Leal Filho, W & Schwarz, J 2008, 'Engaging stakeholders in a sustainability context: the regional centre of expertise on education for sustainable development in Hamburg and region', *International Journal of Sustainability in Higher Education*, vol. 9, no. 4, pp. 498–508.

Leal Filho, W 2011, 'Applied sustainable development. A way forward in promoting sustainable development in higher education institutions', in W Leal Filho (ed.), *World trends in education for sustainable development*, Peter Lang, Frankfurt a.M., pp. 11–30.

Levi-Faur, D 2012, *Oxford handbook of governance*, OUP, Oxford, pp. 4, 20, 21.

Longman, M 2014, 'Aboriginography. A new decolonized aboriginal methodology', in G Guttorm & SR Somby (eds.), *Diedut*, vol. 3, Sami University College, Alta, pp. 16, 17.

Magni, G 2016, 'Indigenous knowledge and implications for the sustainable development agenda. Education for people and planet: creating sustainable futures for all', Background paper prepared for the 2016 Global Education Monitoring Report, UNESCO, p. 5, viewed, https://pdfs.semanticscholar.org/46cc/5a5297743f31abfacce0c8ff89754bdbfe18.pdf.

Marsik, T & Wiltse, N 2018, 'A low carbon Arctic energy system? Challenges, opportunities and trends', in GM Arruda (ed.), *Renewable energy for the Arctic. New perspectives*, Routledge, Oxford, pp. 3–24.

Martínez-Cobo, J 1982, *Study of the problem of discrimination against Indigenous populations*, Final Report, Chapter V, E/CN.4/Sub.2/1982/Add.6, UNDESA, New York, p. 3.

Mathis, JT, Cooley, SR, Lucey, N, Colt, S, Ekstrom, J, Hurst, T, Hauri, C, Evans, W, Cross, JN & Feely, RA 2015, 'Ocean acidification risk assessment for Alaska's fishery sector', Progress in Oceanography, p. 72, viewed 29 November 2016, doi:10.1016/j.pocean.2014.07.001.

Mayr, E 1991, *One long argument: Charles Darwin and the genesis of modern evolutionary thought*, Harvard University Press, Cambridge, MA, p. 97.

McCann, H 2013, 'Local and traditional knowledge stewardship: managing data and information from the Arctic', National Snow and Ice Data Center, viewed 19 November 2016, www.arcus.org/witness-the-arctic/2013/2/article/19956.

Meehan, RH 1995, 'Long-term conservation strategies. Alaska's North Slope oil and gas experience', in RMM Crawford (eds.), *Disturbance and recovery in Arctic lands. An ecological perspective*, vol. 25, Springer, London, Series 2, pp. 519–529.

Mezirow, J 1997, 'Transformative learning: theory to practice', *New Directions for Adult and Continuing Education*, no. 74, pp. 5–12, doi:10.1002/ace.7401.

Michelsen, G 2011, 'Future challenges in the context of sustainable development from a European point of view', in M Barth, M Riechmann & ZA Sanusi (eds.), *Higher education for sustainable development. Looking back and moving forward*, VAS-Verlag für akamische Schriften, Bad Homburg, pp. 59–77.

Michelsen, G 2015, 'Policy, politics and polity in higher education for sustainable development', in M Barth, G Michelsen, M Rieckmann & I Thomas (eds.), *Routledge handbook of higher education for sustainable development*, Routledge, Abingdon, pp. 40–55.

Mulà, I, Tilbury, D, Ryan, A, Mader, M, Dlouhá, J, Mader, C, Benayas, J, Dlouhý, J & Alba, D 2017, 'Catalysing change in higher education for sustainable development: a review of professional development initiatives for university educators', *International Journal of Sustainability in Higher Education*, vol. 18, no. 5, pp. 798–820.

Nanda, P & Pring, G 2013, *International environmental law & policy for the 21st century*, 2nd edn, Martinus Nijhoff Publishers, Leiden, Boston, MA, pp. 283–302.

Natural Resources Canada 2018, *Energy fact book 2018–2019*, Government of Canada, Ottawa.

NETL (National Energy Technology Laboratory) 2017, 'Modern grid initiative', US Department of Energy, viewed 1 September 2017, www.netl.doe.gov/moderngrid/opportunity/vision_technologies.htmlArchived.

Nordic Council of Ministers 2015, *Local knowledge and resource management. On the use of indigenous and local knowledge to document and manage natural resources in the Arctic*, Norden, Norway.

Norwegian Government 2017, *Norway's Arctic strategy – between geopolitics and social development*, Norwegian Ministry of Foreign Affairs and Norwegian Ministry of Local Government and Modernisation, Oslo.

NTET (Nurturing Talent for Energy Technology Program) 2013, Background and goals, www.energyedu.tw/eng/background.php (accessed 20 August 2013).

NTI (Nunavut Tunngavik Inc) 2005, 'What if the winter doesn't come? Inuit perspectives on climate change adaptation challenges in Nunavut', Summary workshop report, 15–17 March 2005, Nunavut Tunngavik, Iqaluit, Nunavut.

Nussbaum, M 2011, *Creating capabilities: the human development approach*, Harvard University Press, Cambridge, MA, pp. 33–34.

Perry, C & Andersen, B 2012, *New strategic dynamics in the Arctic Region. Implications for National Security and International Collaboration*, The Institute for Foreign Policy Analysis, Cambridge, MA, p. 14.

Petrov, A 2014, 'Extensive plan of activities for the Arctic-Frost research network', *Witness the Arctic*, vol. 18, no. 2, pp. 14, 15.

Petrov, A, BurnSilver, S, Chapin, F, Fondahl, G, Graybill, J, Keil, K, Nilsson, A, Riedlsperger, R & Schweitzer, P 2016, 'Arctic sustainability research: toward a new agenda', *Polar Geography*, vol. 39, no. 3, pp. 165–178.

Poppel, B 2015, *SLiCA: Arctic living conditions – living conditions and quality of life among Inuit, Sami and indigenous peoples of Chukotka and the Kola Peninsula*, Nordic Council of Ministers, Denmark, p. 67.

Prime Minister's Office 2015, 'Growth from the North. How can Norway, Sweden and Finland achieve sustainable growth in the Scandinavian Arctic?', Report of an independent expert group, Prime Ministers' Office Publications 04/2015, Prime Minister's Office, Edita Prima, Helsinki, pp. 13, 14, 15.

Pulsifer, P, Gearheard, S, Huntington, HP, Parsons, MA, McNeave, C & McCann, HS 2012, 'The role of data management in engaging communities in Arctic research: overview of the exchange for local observations and knowledge of the Arctic (ELOKA)', *Polar Geography*, vol. 35, no. 3–4, pp. 271–290.

REAP-Renewable Energy Atlas Project 2016, *Renewable energy Atlas of Alaska. A guide to Alaska's clean, local and inexhaustible energy resources*, Alaska Energy Authority, Anchorage, AK.

Rees, WG & Williams, M 1995, 'Satellite remote sensing of the impact of industrial pollution on tundra biodiversity', in RMM Crawford (ed.), *Disturbance and recovery in Arctic Lands. An ecological perspective*, vol. 25, Springer, London, Series 2, pp. 253–282.

Ryan, A 2011, 'ESD and Holistic curriculum change', pp. 3, 5, viewed 12 December 2016, www.heacademy.ac.uk/resources/detail/sustainability/esd_ryan_holistic.

Scollon, R 1981, *Human knowledge and the institutions knowledge: communication patterns and retention in a public university*, Center for Cross-Cultural Studies, University of Alaska, Fairbanks, AK.

Scott, WAH & Gough, SR 2003, *Sustainable development and learning: framing the issues*, RoutledgeFalmer, London, pp. 113, 116.

Sen, A 1993, 'Capability and well-being', in M Nussbaum & A Sen (eds.), *The quality of life*, Clarendon Press, Oxford, pp. 30–53.

Sen, A 2004, *UN human development report 2004: chapter 1 cultural liberty and human development*, UN Human Development Reports, United Nations Development Programme, p. 16, viewed 10 December 2016, http://hdr.undp.org/sites/default/files/reports/265/hdr_2004_complete.pdf.

Senate of Canada 2014, 'Standing committee on energy, the environment and natural resources', Evidence, 2nd Session, 41st Parliament, 29 April, 2014 (Catherine Conrad, Director, Environment and Renewable Resources, Aboriginal Affairs and Northern Development Canada), p. 9.

Senate of Canada 2015, 'Standing Senate Committee on Energy, the Environment and Natural Resources. Powering Canada's Territories', p. 9, viewed http://senate-senat.ca/enev.asp.

Sillitoe, P 1998, 'The development of indigenous knowledge: a new applied anthropology', *Current Anthropology*, vol. 39, no. 2, pp. 223–252.

Smit, B & Wandel, J 2006, 'Adaptation, adaptive capacity, and vulnerability', *Global Environmental Change*, vol. 16, pp. 282–292.

Sterling, S 2001, *Sustainable education: re-visioning learning and change*, Schumacher Briefing, Green Books, Darlington.

Sterling, S 2004, 'An analysis of the development of sustainability education internationally: Evolution, interpretation and transformative potential', in J Blewitt and C Cullingford (eds.), *The Sustainability Curriculum: The Challenge for Higher Education*, Earthscan, London.

Sterling, S. (ed.) 2008, *Sowing seeds: how to make your modules a bit more sustainability oriented: A help guide to writing and modifying modules to incorporate sustainability principles*. Centre for Sustainable Futures, Plymouth University, Plymouth, viewed http://csf.plymouth.ac.uk/files/Sowing%20Seeds%2013%20June%202008.pdf (Accessed March 15th, 2019).

Sterling, S, Irving, D, Maiteny, P & Salter, J 2005, *Linking thinking: new perspectives on thinking and learning for sustainability*, Perthshire, WWF Scotland.

Stern, NH 2007, *The economics of climate change: the stern review*, Cambridge University Press, Cambridge, UK, pp. 20, 65, 68.

Stibbe, A 2009, *The handbook of sustainability literacy: skills for a changing world*, Green Books, Totnes.

Stohl, A, Klimont, Z, Eckhardt, S, Kupiainen, K, Shevchenko, VP, Kopeikin, VM & Novigatsky, AN 2013, 'Black carbon in the Arctic: the underestimated role of gas flaring and residential combustion emissions', *Atmospheric Chemistry and Physics*, vol. 13, p. 8834. doi:10.5194/acp-13-8833-2013.

Sundaresan, S & Bavle, S 2008, 'Student participation and engagement in sustainable human development: a value education approach', in W Leal Filho, L Brandli, P Castro & J Newman (eds.), *Handbook of theory and practice of sustainable development in higher education*, vol. 1, Springer, Hamburg, pp. 171–186.

Sustainable Cleveland 2019 Summit 2009, Action and resources guide. Building an economic engine to empower a green city on a blue lake, Cleveland, p. 9, viewed www.gcbl.org/files/resources/sc2019executivesummary9sep10.pdf

The Nature Conservancy and Wainwright Traditional Council 2008, *Wainwright traditional use area conservation plan map book*, The Nature Conservancy, Anchorage, AK, p. 40.

The Shell Report 1998, *Profits and principles – does there have to be a choice?* Group External Affairs Shell International Shell Centre, Shell International Limited, London, pp. 4, 53, 54, viewed 30 December 2016, shell-sustainability-report-1998-1997.pdf.

Tilbury, D, Keogh, A, Leighton, A & Kent, J 2005, 'A national review of environmental education and its contribution to sustainability in Australia: further and higher education', Report prepared by Australian Research Institute in Education for Sustainability (ARIES) for the Department of the Environment and Heritage, Australian Government, Sydney, p. 1, viewed, www.aries.mq.edu.au/project.htm.

UN General Assembly 2015, 'Transforming our world: the 2030 Agenda for sustainable development', 21 October 2015, A/RES/70/1, p. 14, viewed 5 January 2019, www.refworld.org/docid/57b6e3e44.html.

UNDP 2001, *Human development report, 2001: promoting linkages*, United Nations Development Programme, Oxford University Press, New York, p. 9.

UNDRIP 2007, 'United Nations declaration on the rights of Indigenous peoples', United Nations Permanent Forum on Indigenous Issues, viewed 18 December 2018, www.un.org/development/desa/indigenouspeoples/wp-content/uploads/sites/19/2018/11/UNDRIP_E_web.pdf.

UNESCO 2005, *United Nations Decade of Education for Sustainable Development (DESD) 2005–2014*, UNESCO, Paris, p. 2.

UNESCO 2014, *Education strategy 2014–2021*, UNESCO, Paris, p. 46, viewed 2 December 2018, www.natcom.gov.jo/sites/default/files/231288e.pdf.

UNESCO 2015, 'Global Citizenship Education: Topics and Learning Objectives'. UNESCO, Paris, p. 15, https://unesdoc.unesco.org/ark:/48223/pf0000232993 (accessed 26 November 2018).

UNESCO 2015a, 'Transforming Our World Literacy for Sustainable Development'. Selected case studies, p. 8, http://www.unesco.org/uil/litbase (accessed 12 January 2017).

United States delegation of the Arctic Council 2016, *Adaptation and resilience in the Arctic: A primer on the Arctic resilience report and the adaptation actions for a changing Arctic report*, Arctic Council Secretariat, Tromsø, Norway.

UNPFII9 (Ninth Session of the UN Permanent Forum on Indigenous Issues) 2010, 'Report on the Ninth Session', 19–30 April 2010, UN Headquarters, New York.

UNWCED 1987, *Report of the world commission on environment and development*, General Assembly Resolution 42/187, 11 December 1987, p. 10, viewed 30 March 2018, www.un.org/documents/ga/res/42/ares42-187.htm.

US Department of Energy 2017, *Energy literacy essential principles and fundamental concepts for energy education*, A Framework for Energy Education for Learners of All Ages, p. 1, viewed 1 December 2017, www.energy.gov/sites/prod/files/2017/07/f35/Energy_Literacy.pdf.

US Department of Energy 2017a, 'Essential principles and fundamental concepts for energy education. A framework for energy education for learners of all ages', p. 13, viewed 10 December 2017, http://energy.gov/eere/education/energy-literacy-essential-principles-and-fundamental-concepts-energy-education.

USGS 2008, 'Circum Arctic resource appraisal: estimates of undiscovered oil and gas north of the Arctic circle', USGS Fact Sheet 2008–3049.

Van Treuren, KW & Gravagne, IA 2008, 'Raising community energy awareness: building an energy display at the Mayborn Museum', *Proceedings of the 2008 ASEE Gulf-Southwest Annual Conference*, The University of New Mexico, Albuquerque, NM, pp. 1–11.

van Voorst, RS 2009, 'I work all the time- he just waits for the animals to come back: social impacts of climate changes: a Greenlandic case study', *Jàmbá. Journal of Disaster Risk Studies*, vol. 2, no. 3, pp. 235–252.

Vare, P & Scott, WAH 2007, 'Learning for a change: exploring the relationship between education and sustainable development', *Journal of Education for Sustainable Development*, vol. 1, no. 2, pp. 3, 4.

Vars, LS 2007, 'Hvorfor bør man og hvordan kan man bevare samenes tradisjonelle kunnskap?', in JT Solbakk (ed.), *Árbevirolašmáhttu ja dahkkivuoigatvuohta – Tradisjonell kunnskap og opphavsrett –Traditional knowledge and copyright*, Sámikopija, Karasjok, pp. 123–166.

Webster, DH & Zibell, W 1970, *Iñupiat Eskimo Dictionary*, Department of Education. University of Alaska, AK.

Wiessner, S 2007, 'The UN Declaration on the rights of indigenous peoples', General Assembly resolution 61/295, 13 September 2007, New York, p. 4.

Young, M 2008, 'From constructivism to realism in the sociology of the curriculum', *Review of Research in Education*, vol. 32, no. 1–32, p. 14.

Young, M 2013, 'Powerful knowledge: an analytically useful concept or just a 'sexy sounding term? A response to John Beck's "powerful knowledge, esoteric knowledge, curriculum knowledge"', *Cambridge Journal of Education*, vol. 43, pp. 195–198.

Young, M 2014, 'Powerful knowledge as a curriculum principle', in M Young, D Lambert, C Roberts & M Roberts (eds.), *Knowledge and the future school: curriculum and social justice*, Bloomsbury Academic, London, pp. 65–88.

3 The influential theories for the Arctic Higher Education – Education for Sustainable Development (ESD)

3.1 Introduction

This chapter presents the main concepts informed by underpinning theories that provide an interactive and integrative foundation for the thinking behind this book and for the models designed and derived from the data. Complexity Theory, linked to Capability Theory, Citizenship Theory, and Sustainable Development Theory are all outlined. Arising from the research questions are three areas of interest. The first concerns the aims within the curricula reflecting energy literacy and UNESCO Education for Sustainable Development ESD; the second concerns practice from educator and student perspectives, including the capabilities that each group perceives are being developed; and finally, the processes through which the curricula are 'made and reviewed' are investigated. This chapter also aims at expanding our understanding of the SD concept and pillars and how energy production/use is linked to well-being, capabilities, and human development in a complex geographical area where cultural aspects are entangled with environmental, economic, and social factors.

3.2 Complexity Theory – the dynamic nature of change

For the last two centuries, the world has been explained according to a mechanized way of thinking based on the industrial revolution and the trend to mechanize and polarize the own process of thinking. This mechanical worldview not well adapted to change 'continues to maintain its attraction as it provides a sense of order, purpose and control' (Boulton, Bowman & Allen 2015, p. 20), but it is not capable of capturing, perceiving, nor explaining the changes and trends.

This mechanical worldview has underpinned processes of management, policymaking, and education by defining their role in the world and shaping how they engage with life and work (Boulton, Bowman & Allen 2015, p. 1). The mechanical perspective assumes that it is possible to control the future of the economy, society, ecology, energy resources, and climate, because this mind-set is based on a stable, non-changeable, non-dynamic worldview. Consequently, it does not lead one to explore changeable multiple perspectives, interrelationships, unpredictability, interdisciplinary knowledge, or co-evolution. To conceptualize Complexity Theory is a complex task and understanding this concept depends on the historical background of the concept of complexity.

'Complexity theory is a theory of change, evolution, adaptation and development for survival' in the definition of Morrison (2008, p. 16). The author continues arguing that:

> It breaks with simple successionist cause-and-effect models, linear predictability, and a reductionist approach to understanding the phenomena, replacing them with organic, non-linear and holistic approaches respectively (Santonus 1998, p. 3), in which relations within interconnected networks are the order of the day.
> (Wheatley 1999, p. 10; Youngblood 1997, p. 27)

Additionally, Mason (2008, p. 33) argues that Complexity Theory:

> concerns itself with environments, organizations or systems that are complex in the sense that very large numbers of constituent elements or agents are connected to and interacting with each other in many different ways. These constituent elements or agents might be atoms, molecules, neurons, human agents, institutions, corporations, etc. Whatever the nature of these constituents, the system is characterized by a continual organization and re-organization of and by these constituents' Mason (2008, p. 33) 'into larger structures through the clash of mutual accommodation and mutual rivalry. Thus, molecules would form cells, neurons would form brains, species would form ecosystems, consumers and corporations would form economies, and so on. At each level, new emergent structures would form and engage in new emergent behaviours. Complexity, in other words, is really a science of emergence.
> (Waldrop 1993, p. 88)

Charles Darwin and his 'The Origin of Species' published in 1859 is one of the precursors of Complexity Theory (also called 'complexity thinking' or 'systems thinking') and adaptability (von Bertalanffy 1969, p. 12), because as Mason (2008, p. 33) argues 'Complexity is inherently systemic in nature'. Darwin's ideas were reflected by the phrase 'survival of the fittest'. 'Survival of the community most fitted to its surroundings' (Spencer 1864, p. 144) is probably a more appropriate terminology to understand Darwin's idea of a 'path-dependent emergence of new forms of life more adapted than what came before'. This is also a point advocated by the physicist Ilya Prigogine (1980) when he referred in his research about 'order and structure' to 'the spontaneous dynamics of living systems' and random circumstances that can lead to the transformation, evolution, or 'the autocatalytic' events generating momentum to a new direction.

Darwin's ideas can be perfectly applied to social sciences as adaptability is a natural dynamic that can be applied to 'systems' regardless of their nature (natural, social, educational, corporative, organizational, etc.) as observed by Waldrop (1993). In this sense Darwin's conclusions (Boulton, Bowman & Allen 2015, p. 63) are relevant to this book because:

(a) variation, understood as fluctuation, was essential for creativity, change and adaptability;
(b) the future builds on and is shaped by the past; and yet
(c) the future is unknowable in advance.

(Boulton et al. 2015, p. 63)

In other words, change is relevant to stimulate creativity and adaptability being Complexity 'a theory of perpetual novelty, disequilibrium and creativity' (Morrison 2008, p. 29) by admitting that knowledge is partial and incomplete; the future depends on historical dynamics, it is unpredictable and constructed by an adaptive re-organization of interactive and interconnected components interpreted according to emergent patterns. These are ideas perfectly applied to Higher Education (HE) because complexity thinking 'troubles the modern and western habit of thinking in terms of discontinuities around matters as theory and practice, knowers and knowledge, self and other, mind and body, art and science, student and curriculum' (Davis 2008, p. 47). It has applicability in relation to HE regarding to educational change (Mason 2008, p. 45) because 'new properties and behaviours' can emerge 'not only from the constituent elements of a system but from the myriad of connections among them'. These are the premises to understand a complex world characterized as systemic and synergistic, multi-scalar, variable, and path-dependent, where new patterns can emerge and with more than one future (Boulton et al. 2015, p. 35). A world that cannot be mapped by mechanical thinking creates the need of perspectives aligned to complexity thinking capabilities.

These perspectives indicate the importance of teaching skills or developing capabilities for complex thinking by adopting a systems approach when covering the course content to stimulate an extended understanding or a holistic perspective of interconnected components that, for this book, are the economic, social, environmental, and cultural interconnections. This is a challenge embraced by the HE capability researchers as Young (2008) and Lambert (2014) whose positions are aligned to a 21st century knowledge based on 'the counterintuitive sense of the planet as a place, with its physical and human interdependencies' (Lambert 2014, p. 19).

This book argues that the complex Arctic is the stage for this new thinking exercise, as it presents a highly synergistic relationship of its species which are susceptible to collapse in the face of change. In other words, the Arctic is an especially sensitive environment with highly specialized species finely co-adapted making it extremely difficult to preserve the current systems (food chain, ecosystems, social systems) and to adapt to variations or changing patterns (Arruda & Krutkowski 2016, p. 519). The interesting phenomenon is that the same behaviour applies to the Arctic social system with its multicultural complexities.

Last but not least, another relevant reason for using Complexity Theory in this book is to usefully tie together the curriculum-making process in the context of Arctic ecosystems, capabilities, and ESD/EL by looking at the particular process of curriculum design related to specific courses in specific Arctic places, as a lens for understanding the more complex system of HE, Arctic change, and Arctic energy production and literacy.

3.2.1 Holistic education for the dynamic nature of change

As seen in Section 3.2, it is not an easy task to conceptualize Complexity Theory or complexity thinking. To unpack these concepts, it was necessary to understand them in the context of systems. In 'Complexity and Flexibility', Moses (2004, p. 2) argues that a system is complex when it is composed of many parts that interconnect in intricate ways. This definition has to do with the number and nature of the interconnections. Metric for intricateness is represented by the amount and nature of information contained in the system. When trying to understand the interconnections of the parts and the whole of intricate systems, flexibility and holism may help in capturing the dynamic nature of change, what counterpoints straightforward to the mechanical perspective of seeing the world.

In this post-Darwinian era, new theories emerged, including Relativity and Quantum Mechanics more associated to a non-reductionist worldview, uncertainty, and possibility (Boulton et al. 2015, p. 66). It was during this time when Smuts (1999) advocated that interconnections through synthesis lead to 'wholes' and that this is a common feature of the evolutionary process. In this sense, Smuts captured the dynamic nature of change: 'As **holism is a process of creative synthesis**, the resulting wholes are not static but dynamic, evolutionary, creative' and holism is the root of systems thinking, where situations need to be explored by looking at interconnections (von Bertalanffy 1969, p. 12). Thinking holistically and understanding systems are capabilities that education in the Arctic needs to develop.

Complexity theory sees change as arising at the microscopic level, in the detailed interactions in particular circumstances, what seems to match perfectly the aims of this study as the Arctic interactions occur at a **macro** (holistic) level and **microscopic** level simultaneously and one level is not less important than the other level for the equilibrium of the Arctic system.

No one seems to have the answer to the question of how to accept a 'world as complex' versus a 'world as machine' stance or how to operate this transition of viewpoint. The idea of merely accepting that the world is complex instead of predictable and controllable will probably offer better guidance and strategies on how to respond to the world's complexity and it will possibly allow the human being to embark in a more sustainable path. The idea of sustainability is naturally interdisciplinary as it is based on interconnected, eco-systemic interrelationships among organisms and species that are interdependent. It involves natural processes of maintaining and sustaining life in an organic planet that cannot be perceived adequately through a mechanic view. It is important to remark that the same process happens at the social level. This shift in the perception has implications in the way we understand knowledge, the processes of change, and the nature of reality; it is exactly what provides a foundation for the qualitative methodology adopted in Chapter 4 of this book. The point is the need for flexibility and creativity on the sense of impermanence and uncertainty when examining any approach and, perhaps, the adoption of a more holistic perspective of realizing how connected personal experiential perspectives are to the scientific view. This latest statement can be largely evidenced by the interlinks suggested in this study between Traditional Knowledge and Westernized scientific knowledge, or the distinct perspective permeating methods and pedagogies of Arctic Indigenous population which proves the point made by Boulton et al. (2015, p. 10) when saying that complexity as a 'new science' has power.

The knowledge derived from complexity has power because it allows the construction of alternative futures and by recognizing that different people have different personal, social, and cultural constructions of reality; this is a first step to understand and interpret these realities and to envisage alternative paths. These possibilities justify the drawing on some theoretical foundations and the use of the constructivist approach in the conceptual framework of this book and the interpretivist methodology to the data analysis comprehending Western and Indigenous perspectives.

3.2.2 Complexity Theory and the Arctic environment – the Arctic as a complex adaptive system

The Arctic presents its contradictions and complexities and it needs to be understood as a system, a complex system. At the same time, it appears robust; it is sensitive and vulnerable to changes and socio-environmental pressures. It is a specialized system formed by physical and human interconnected dimensions. The physical sub-systems are formed by glaciers, areas of tundra, mountains, fjords, and ocean. The human sub-system is formed by communities and their economic, social, and cultural activities. In relation to the former or the latter, one change can affect the whole system. The physical, chemical, biological, and social components of the

Arctic System are interrelated, and, therefore, a holistic perspective is needed to understand and quantify their connections and predict future system changes (Roberts et al. 2010, p. 3). This book explores whether this is the case in the Arctic curriculum.

The Arctic with its complexity is increasing in perceived importance as an area to conduct and apply the findings of research to policy and to practice, in the field. The core focus of the proposed Arctic System Model (ASM), proposed by Roberts et al. (2010, p. 3), is to understand complexity and adaptation in the Arctic System as well as society's role and response in the evolution of that system through modelling and observations. The modelling programme aims at quantifying and reducing uncertainties related to geo-biophysical and social variability of the Arctic System, uncertainty in the models themselves, and uncertainty in society's response and adaptation to Arctic change.

The Arctic System is the northern dome of the geosphere and biosphere that circulates energy, mass, and nutrients between areas inside the Arctic Circle and the mid-latitudes (Roberts et al. 2010, p. 3). This definition is physically based and refers to the relevant terrestrial and marine ecosystems that are integral to the Arctic environment. It sets the area for the Arctic System modelling and measurement; however, the social system is still not easy to measure. In other words, the air and sea temperatures, isotherms, and ice cover are as important as the well-being of its peoples. This more integrated perspective may promote further understanding of Arctic change and its consequences for socio-environmental local and global systems. The adaptation of the system occurs both in the field (physical) and in conceptualization (theoretical framework).

3.2.3 The influence of Complexity Theory on knowledge systems and curriculum

Complexity Theory, as a theory of change, development, and evolution through relationships, raises an interesting agenda for education because for Fullan (1989) change equals learning. Learning is a central element in both complexity theory and education.

In complex and changeable contexts learning occurs, not through direct transmission from expert to novice, but in a non-linear manner through all in a class exploring a problem together. In other words, the curriculum (with its expression in a syllabus) emerges within an ongoing process that catalyzes itself via interactions within the system or network. Consequently, the aim of embracing complexity according to Doll's (1993, p. 11) postmodern perspective is:

> a process of cross-fertilization, pollination, catalyzation of ideas: Over time (...) a network of connections and interconnections becomes more and more webbed
>
> (Doll 1993, p. 11)

Education for sustainable development   85

According to this view, 'knowledge' and 'reality' should not be understood as separate systems, but as part of the same emerging complex system. Boulton, Bowman and Allen (2015, p. 11) reinforces this view by stating that:

> Complexity provides an ontology, a worldview, a genetic insight into dynamics, into the way the world becomes and allows exploration and insight into the particularity of problems and areas of interest.
>
> (Boulton et al. 2015, p. 11)

This perspective allows for Indigenous views and Western views to be given equal power/currency because reality is a construction and there is no right nor wrong answer, but different personal, social, and cultural constructions will be the outcome as mind constructs reality.

When considering the purpose of education in terms of an emergent understanding of knowledge and reality, it is possible to envisage education as a practice which makes possible a dynamic, self-renewing, and creative engagement with 'content' or 'curriculum' by means of which learners are able to respond and hence bring forth new worlds (Osberg & Biesta 2007).

With this conception of knowledge, the purpose of the curriculum is no longer to facilitate the acquisition of knowledge about reality. This 'epistemology of emergence' (Osberg et al. 2008, p. 213) therefore calls for a shift in curricular thinking, away from questions about presentation and representation and towards questions about engagement and response. This means questions about what to present in the curriculum and whether these things should be directly presented or should be represented are no longer relevant as curricular questions. While content is important, the curriculum is less concerned with what content is presented and how, and more with the idea of the extent to which content is engaged with and responded to (Mason 2008, p. 12).

> With this conception of knowledge and the world the curriculum becomes a tool for the emergence of new worlds, rather than a tool for stabilization and replication.
>
> (Mason 2008, p. 12)

This is the knowledge that leads to consciousness, but to develop this level of engagement and action, education needs to develop students' capabilities.

3.2.4 Powerful knowledge – the origin of geo-capabilities

The interdisciplinary knowledge-base and geo-cultural perspective needed to understand the Arctic complexity under a systemic and holistic viewpoint seem to be supported by the components of what Young (2008, p. 14) refers to as 'powerful knowledge':

> Powerful knowledge refers to what the knowledge can do or what intellectual power it gives to those who have access to it. Powerful knowledge provides more reliable explanations and new ways of thinking about the world and (...) can provide learners with a language for engaging in political, moral, and other kinds of debates.
>
> (Young 2008, p. 14)

The reason this knowledge is powerful resides in the fact that 'it provides the best understanding of the natural and social worlds' (Young 2013, p. 196) by emphasizing the interconnections among elements and it also enables people to envisage alternatives (Young 2014, p. 74) or futures scenarios. It opens possibilities, new ways of thinking about the world, and consequently analyzing, understanding, and participating in debates on significant local, national, and global issues, as it is systemic and specialized (Maude 2015, p. 20). This is the knowledge that can enable students to become active citizens in the complex modern world (Young 2011). It has the power to reframe science and knowledge by paying attention to the local detail, or the geographical perspective of complexity, but at the same time to have on board the global perspective. Powerful knowledge expresses the link between complexity and geo-capabilities being a relevant definition, to envisage the large-scale processes of change the Arctic presents and other realities that can put this concept in checkmate.

3.3 Capability Theory

More recent understandings of development emphasize the use and management of natural resources to satisfy human needs and improve people's quality of life. They also add concerns for people's health and education as key components to create more dynamic economies and higher material prosperity for societies. Additionally, others view development as enabling people to live lives they value; however, it is important not to forget that values are different in different cultures.

The different assumptions about what constitutes development have important policy and regulatory implications as well as practical consequences for people's lived realities. Development has been seen for long as economic growth, as per the post war view of expansion and reconstruction. This focus on annual economic growth and quantity predominates today over the idea of quality of growth and the implications of this quality on people's lives. Economic growth is remarkably a necessary component of development; however, it continues, nowadays, to be the predominant framework for development permeating and dominating the SD discourse. The negative side of this phenomenon is the fact that governments justify their measures based on economic growth not considering the important balance that SD embeds.

One approach to development is the one in which the objective is to expand what people are able to do and be or their real freedoms. This is a human-centred view of development where a healthy economy is one that enables people to enjoy a long and healthy life, a good education, a meaningful job, physical safety, democratic debate, and so on. According to this perspective, the analysis shifts from the economy to the person. The currency of assessment shifts from money to the things people can be and can do in their lives now and in the future. This is the view developed by the philosopher and Nobel laureate in economics Amartya Sen, whose writings on the 'capability approach' explain what capabilities are and provide the philosophical basis of human development. Under Sen's perspective, income is obviously an important instrument in enabling people to realize their full potential despite not being the most important. Consequently, the basic premise for development is to enlarge people's choices and consequently their freedom (Alkire & Deneulin 2009, p. 35).

For Haq (2004, p. 19) the purpose of development is to enlarge all human choices and not just income. The human development paradigm is concerned both with building human capabilities (through investment in people) and with using those human capabilities more fully (through an enabling framework for growth and employment). Human development has four essential pillars: equality, sustainability, productivity, and empowerment (Haq 2004, p. 19). It regards economic growth as essential, but emphasizes the need for its quality and distribution, and its link with human lives and long-term sustainability. The human development paradigm defines the ends of development and analyzes sensible options for achieving them.

The first Human Development Report in 1990 defined human development as 'both the process of widening people's choices and the level of their achieved well-being' (UNDP 1990, p. 9). The perspective of human development incorporates the need to remove the obstacles that people face through the efforts and initiatives of people themselves, through their own capabilities. This effort towards the collective well-being and freedom also improves societal organization and commitment. These are the two central ideas that give cogency to the focus on human development according to Amartya Sen's view (Fukuda-Parr & Kumar 2003, p. 7).

The richness of this perspective points out to the human development approach as being inherently multi-dimensional and plural. In practice, most policies focus on one or several components of human development, the approach itself is potentially broad. It is about education as much as it is about health. It is about culture as much as it is about political participation (Alkire & Deneulin 2009, p. 28).

The human development approach inspired by Amartya Sen's pioneering reflection in welfare economics, social economics, and development economics has provided the basis of a new paradigm in economics and in the social sciences related to the capability approach. Amartya Sen writes:

> A person's capability to achieve 'functionings' that he or she has reason to value provides a general approach to the evaluation of social arrangements, and this yields a particular way of viewing the assessment of equality and inequality.
>
> (Sen 1992, p. 5)

The key idea of the capability approach is that social arrangements should aim to expand people's capabilities – their freedom to promote or achieve what they value doing and being. An essential test of development is whether people have greater freedoms today than they did in the past. A test of inequality is whether people's capability sets are equal or unequal.

In terms of the capability approach, Sen (1999, p. 75) proceeds defining 'functionings' as 'the various things a person may value doing or being' (Sen 1999, p. 75). In other words, 'functionings' are valuable activities and states that make up people's well-being – such as being healthy and well-nourished, being safe, being educated, having a good job, being able to visit loved ones. They are also related to goods and income but describe what a person can do or be with these. For example, when people's basic need for food is met, they enjoy the functioning of being well-nourished. Capability refers to the freedom to enjoy various functionings. According to Sen's view, capability is defined as 'the various combinations of functionings (beings and doings) that the person can achieve. Capability is, thus, a set of vectors of functionings, reflecting the person's freedom to lead one type of life or another (...) to choose from possible livings' (Sen 1992, p. 40). Put differently, capabilities are, 'the substantive freedoms [a person] enjoys leading the kind of life he or she has reason to value' (Sen 1999, p. 87). Finally, Sen defines 'agency' as the ability to pursue goals that one values and has reason to value. An agent is 'someone who acts and brings about change' (Sen 1999, p. 19).

The capability approach refers to the freedom to achieve well-being as a matter of what people are able to do and to be, and the kind of life they are effectively able to lead. In this sense, it is generally conceived as a flexible and multi-purpose framework, rather than a precise theory of well-being (Robeyns 2005, pp. 94, 96; Sen 1992, p. 48). Capabilities are a person's real freedoms or opportunities to achieve functionings.

Functionings are 'beings and doings', or the various states of human beings and activities that a person can undertake. Functionings related to 'beings' are, for instance, being well-nourished, being undernourished, being housed, being educated, being illiterate, being literate, being healthy, or being depressed. Examples of functionings related to 'doings' are taking part in a debate, voting in an election, travelling, killing animals, eating animals, consuming fuel in order to heat one's house, and donating money to charity (Robeyns 2016, p. 3).

Functionings are constitutive of a person's being, and an evaluation of well-being has to take the form of an assessment of these constituent elements (Sen 1992, p. 39). Sen continues saying that:

> Whereas 'functionings' are the proposed conceptualization for interpersonal comparisons of (achieved) well-being, 'capabilities' are the conceptualization for interpersonal comparisons of the freedom to pursue well-being, which Sen calls 'well-being freedom'.
> (Sen 1992, p. 40)

Nussbaum (2000, p. 72) moves from theory to empirical applications when endorsing a specific list of capabilities. She justifies this list by arguing that:

> Each of these capabilities is needed in order for a human life to be 'not so impoverished that it is not worthy of the dignity of a human being'. These capabilities are the moral entitlements of every human being on earth.
> (Nussbaum 2000, p. 72)

Nussbaum (2011, p. 33) identified core capabilities (see Table 3.1) as opportunities to experience a dignified lifespan, as follows:

Table 3.1 Core capabilities according to Nussbaum (2011).

1. *Life. Being able to live to the end of a human life of normal length; not dying prematurely, or before one's life is so reduced as to be not worth living.*
2. *Bodily Health. Being able to have good health, including reproductive health; to be adequately nourished; to have adequate shelter.*
3. *Bodily Integrity. Being able to move freely from place to place; to be secure against violent assault, including sexual assault and domestic violence; having opportunities for sexual satisfaction and for choice in matters of reproduction.*
4. *Senses, Imagination, and Thought. Being able to use the senses, to imagine, think, and reason – and to do these things in a 'truly human' way, a way informed and cultivated by an adequate education, including, but by no means limited to, literacy and basic mathematical and scientific training. Being able to use imagination and thought in connection with experiencing and producing works and events of one's own choice, religious, literary, musical, and so forth. Being able to use one's mind in ways protected by guarantees of freedom of expression with respect to both political and artistic speech, and freedom of religious exercise. Being able to have pleasurable experiences and to avoid non-beneficial pain.*
5. *Emotions. Being able to have attachments to things and people outside ourselves; to love those who love and care for us, to grieve at their absence; in general, to love, to grieve, to experience longing, gratitude, and justified anger. Not having one's emotional development blighted by fear and anxiety. (Supporting this capability means*

(*Continued*)

Table 3.1 (Cont.)

---

supporting forms of human association that can be shown to be crucial in their development.)
6. *Practical Reason.* Being able to form a conception of the good and to engage in critical reflection about the planning of one's life. (This entails protection for the liberty of conscience and religious observance.)
7. *Affiliation.*
    a) Being able to live with and toward others, to recognize and show concern for other humans, to engage in various forms of social interaction; to be able to imagine the situation of another. (Protecting this capability means protecting institutions that constitute and nourish such forms of affiliation, and also protecting the freedom of assembly and political speech.)
    b) Having the social bases of self-respect and non-humiliation; being able to be treated as a dignified being whose worth is equal to that of others. This entails provisions of non-discrimination on the basis of race, sex, sexual orientation, ethnicity, caste, religion, national origin and species.
8. *Other Species.* Being able to live with concern for and in relation to animals, plants, and the world of nature.
9. *Play.* Being able to laugh, to play, to enjoy recreational activities.
10. *Control over one's Environment.*
    a) Political. Being able to participate effectively in political choices that govern one's life; having the right of political participation, protections of free speech and association.
    b) Material. Being able to hold property (both land and movable goods), and having property rights on an equal basis with others; having the right to seek employment on an equal basis with others; having the freedom from unwarranted search and seizure. In work, being able to work as a human, exercising practical reason and entering into meaningful relationships of mutual recognition with other workers.

Source: Nussbaum 2011, p. 33

In order to achieve their potential, people need to be able to stay healthy and take part in cultural, economic, social, and political life. Broadly speaking, they need to be in a position to take responsibility for their lives. They need to be able to think, make decisions, and act according to what they believe is right. All these abilities, capacities, and attributes we refer to as human 'capabilities' (Nussbaum 2006, p. 78).

Applying a geographical approach to Nussbaum's capabilities, a framework on how geography can contribute to the development of people's intellectual functioning emerged as 'geo-capabilities' related to making choices for sustainability, being creative and productive in a global economy and culture, and achieving personal autonomy (Solem et al. 2013, p. 216):

> A capabilities approach to geography education asks teachers to consider the role of geography in helping young people reach their full human potential. Geography does not tell us how to live; but

thinking geographically and developing our innate geographical imaginations can provide the intellectual means for visioning ourselves on planet earth.

(Wadley 2008, p. 650)

'Geo-capabilities' can be defined as the educational approach through which an individual can develop a greater potential to lead a life that she or he has reason to value, if they acquire geographical knowledge, enabling them to think geographically (Solem et al. 2013).

Geo-capabilities are focused on those capabilities in Nussbaum's list pertaining to human cognitive abilities and intellectual development, and then phrased in a manner that enables analysis of the curricular role of geography in helping young people think about their life in relation to themselves in the world and what may become of their communities as well as people, places, and environments around the world. The geo-capabilities project in education (initiated at school level geography and applied to HE) makes the explicit claim that the capabilities approach will enable and facilitate international communication about geography in education (Lambert et al. 2015, p. 217; Walkington et al. 2018, p. 11). They ensure the development of a progressive knowledge-led curriculum by seeking a different approach based on different perspectives. For curriculum design, this implies thinking about the role of geographic knowledge, skills, perspectives, and values in developing the capabilities of young people. It also implies thinking in terms of how young people may become deprived of certain capabilities when they lack access to the powerful knowledge provided by geography education (Donert 2015, p. 2; Solem et al. 2013, p. 219).

This perspective seems to be aligned to the ideas that curricula which only focus on competencies for paid employment are deficient (Marsh 2009, p. 7), whereas capabilities represent more than only skills and knowledge. The developmental model of curriculum planning (Kelly 2009, p. 99) suggests that educationists should consider particular views of humanity and human development, including social development and cultural components. In Kelly's words, 'it sees the individual as an active being, who is entitled to control over his or her destiny and sees education as a process by which the degree of control available to each individual can be maximized' or, in other words, 'the central concern is the individual empowerment' (Kelly 2009, p. 99). Consequently, the curriculum is more than stored knowledge, but it should be 'activity', 'experience' (Board of Education 1931, para.75), a source of autonomy. Autonomy, according to Kelly (2009, p. 99) not only means 'freedom from constraints', but also the ability to develop capacities that will enable the individual to make personal choices, decisions, and judgements as an expression of a genuine control over his or her destiny. This notion applied to geographical knowledge and the capacity of exercising this autonomy where he or she lives, provides a powerful tool to live in the Arctic. This integral geographical perspective may equip students in Higher Education for the challenges of the 21st century, among them the challenges of a changing Arctic, being transformed by energy production and use (Boni & Walker 2013, p. 20). This argument reinforces Ryan's idea (Ryan 2011, p. 5) when arguing that a broad concept of 'holistic' curriculum change can be used to guide the review in its inclusive approach to these educational approaches that engage the entirety of the human personality and promote connectivity with the natural world. Practical and conceptual criteria can be established to set the boundaries and scope for the review, with both explicit and implicit links to ESD.

Interdisciplinary participatory pedagogies, 'real world' research, and a systemic view of sustainability are required to meet the transformative aim of Higher Education (Tilbury 2011, p. 3). Energy literacy, as a component of Sustainable Development Literacy (or eco-literacy), requires understanding of complex systems (Complexity Theory is employed to integrate human and physical aspects as outlined in Section 3.2 (Dale & Newman 2005, p. 354), operating effects in cognitive (knowledge), affective (attitudes and values), and behavioural domains.

3.3.1 The relationship between geo-capabilities and energy literacy

Energy literacy seems to have an important role in this process of enhancing understanding about transition among stakeholders and enabling individuals and communities to make well-informed decisions. There are studies carried out by a few scholars in relation to the importance of improving energy literacy in HE not only to enable people to exercise ESD but to be able to make educated energy decisions to contribute to SD. For instance, van der Horst et al. (2016), pointed out the importance and the need to improve 'energy literacy', equipping people to make more thoughtful, responsible energy-related decisions and actions.

There are pioneering studies covering this topic (see Section 2.8) and there is no definite or final model to assess energy literacy levels in HE courses. DeWaters and Powers (2011) conceptualize energy literacy in relation to secondary school, as comprising three domains: cognitive (knowledge), affective (attitudes and values), and behavioural (DeWaters & Powers 2011, p. 1700).

To design an energy literacy pedagogical model, however, it is important to consider concepts and principles of energy literacy. Any model to be proposed depends primarily on these fundamental components that can be found in the definition of an energy literate person that provides a more comprehensive concept, as follows:

(...) one who has a sound conceptual knowledge base as well as a thorough understanding of how energy is used in everyday life, understands the impact that energy production and consumption have on all spheres of our environment and society, is sympathetic to the need for energy conservation and the need to develop alternatives to fossil fuel-based energy resources, is cognizant of the impact that personal energy-related decisions and actions have on the global community, and – most importantly – strives to make choices and exhibit behaviors that reflect these attitudes with respect to energy resource development and energy consumption.

(DeWaters & Powers 2011, p. 1700)

From this concept, it is possible to transpose the following characteristics into a possible pedagogical model: by showing the pedagogical foundations of the concept.

Table 3.2 Pedagogical model designed based on DeWaters and Powers (2011, p. 1700)

Energy Literacy	Pedagogical model as per DeWaters and Powers (2011, p. 1700)
Cognitive	To have conceptual knowledge about energy
Cognitive	To have meaningful understanding of the everyday life energy use
Cognitive	To have meaningful understanding of the impact of energy production/use on environment and society
Cognitive	To have meaningful understanding of the impact of personal educated energy decisions and actions on global community
Cognitive, attitudes, and values	To have meaningful understanding of the needs for: • Energy conservation • Alternatives to fossil fuels
Attitudes, values, and behaviour	To make choices and exhibit behaviours reflecting attitudes according to this understanding concerning energy resource development and consumption

In this process of structuring a more updated and contemporary energy literacy model, by adding to the previous model the components from Hein (1991), Mezirow (2009), and St. Clair (2003), we achieve a more expanded model in terms of pedagogical foundations, focusing also on capability(ies) as the ability of a person to achieve their objectives meaning enactment, doing, and being (Hinchliffe & Terzi 2009; Nussbaum 2011):

Table 3.3 DeWaters and Powers (2011) pedagogical model adapted by foundations from St. Clair (2003), Hein (1991), and Mezirow (2009)

Energy Literacy	Pedagogical model based on DeWaters & Powers, 2011, p. 1700, with additional pedagogical foundations from St. Clair (2003), Hein (1991), and Mezirow (2009)
Cognitive	To have conceptual knowledge about energy
Cognitive	To have meaningful understanding of the everyday life energy use
Cognitive	To have meaningful understanding of the impact of energy production/use on environment and society
Cognitive	To have meaningful understanding of the impact of personal educated energy decisions and actions on global community
Cognitive, attitudes, and values	To have meaningful understanding of the needs for: • Energy conservation • Alternatives to fossil fuels
Capability – critical thinking informing educated energy decisions	To have the capacity to assimilate and critique information to inform choices, attitudes, and sustainable behaviours (St. Clair 2003)
Capability Transformative learning	To reflect upon problematic perspectives and mindsets (Mezirow 2009)
Attitudes, values, and behaviour	To make choices and exhibit behaviours reflecting attitudes according to this understanding concerning energy resource development and consumption
Applications to real life (constructive learning)	Where a route from current knowledge to new information is paved through applications to real life (Hein 1991)

Capabilities related to geographical contexts or 'geo-capabilities' represent an important element of adaptation for the HE curriculum in preparing students for life and work in the complex modern era of globalization and interdependence (Walkington et al. 2018, p. 11). They represent more than skills to be acquired but thinking and learning for lifelong development emphasizing the special value of contextual geographical knowledge for autonomy and freedom and for envisioning alternative futures (Elkind 2004, p. 307; Vygotsky 1978, p. 57) for people, places, and environments (Lambert et al. 2015).

3.4 Educational Theory of Global Citizenship – Global Citizenship Education (GCE)

SD requires changes in energy systems, systems of production and consumption, but also new ways of viewing the world. To change the ways of viewing the world, it is necessary to develop new ways of thinking, acting, and behaving. It implies new levels of responsibility informing values, skills, and knowledge. It also implies not only the capacity of changing the world but reshaping ourselves to live in the world.

Global citizenship, in general terms, means a sense of belonging to a broader community, beyond national boundaries, emphasizing our common humanity and the interconnectedness between peoples as well as between the local and the global levels. It is based on the universal values of human rights, democracy, non-discrimination, and diversity and consists of voluntary practices oriented to social justice and global consciousness (UNESCO 2016, p. 6).

Global citizenship is fundamentally aligned to three specific ideas:

a) Learning to live together, self-identification with the whole of humanity, developing emotional intelligence (compassion), and intercultural understanding to interact constructively across cultural boundaries (Haigh 2014, p. 14);
b) Eco-literacy (learning to live together sustainably) to operate within the limits of the planet (Haigh 2014, p. 14);
c) Responsibility, ethics, fairness, and equity (Haigh 2014, p. 14).

The self-identification with the whole of humanity refers to a sense of belonging to a broader community, beyond the boundaries of a national identity, a tribe or a nation, but referring to a planetary citizen, by understanding the interconnectedness and the interdependency of the beings living on planet Earth with a clear commitment to the collective good (UNESCO 2014b, p. 14).

Eco-literacy is an educational paradigm expressing the ability to understand the natural systems through an integrated approach to deal with environmental problems related to the organization of ecosystems, energy systems, and social systems (UNESCO 2014a). It is ecological intelligence allowing to comprehend and navigate the complexities between the natural and human systems' interaction (Goleman 2009). The attribute of an ecologically literate person consists in learning to think about the world within a holistic, complex, systemic, and sustainable manner. Energy literacy is a fundamental aspect of eco-literacy because energy is in the genesis of the life processes.

Global citizenship implies a higher level of responsibility, ethics, fairness, and equity in attitudes and decision-making (Haigh 2014, p. 14). Responsibility is embedded in global citizenship; it involves living responsibly by understanding that the effects of choices and actions taken in one place will certainly affect every level of life, family, community, and the whole world (Haigh 2014, p. 16).

Theories of global citizenship involve at least three approaches related to Political Theory of Global Citizenship, Educational Theory of Global Citizenship, and the Social Theory of Global Citizenship. The next section addresses the theoretical approach of this book that is the Educational Theory of Global Citizenship or Global Citizenship Education (GCE) that is represented by the components of the UNESCO Global Citizen Education Framework.

3.4.1 Global Citizenship Education (GCE) and Education for Sustainable Development (ESD)

GCE and ESD pursue the same vision of empowering students of all ages when providing a means of understanding the interconnected elements of the world and the complexities of the global challenges faced by humanity in order to develop 'global consciousness' and 'global competence' (Dill 2013). They focus on the following components:

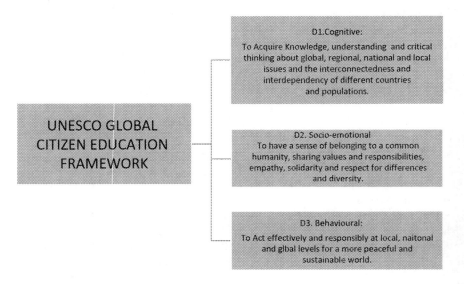

Figure 3.1 UNESCO Global Citizen Education Framework. Core dimensions of Global Citizenship.
Source: Pittman (2017)

Global consciousness and competence involve the development of cognitive, socio-emotional, and behavioural aspects that influence education in terms of contents, outcomes, and learning process. Values, change, and transformation are in the centre of Global Citizenship Education (GCE) as per Table 3.4.

Table 3.4 Global Citizenship Education: topics and learning objectives.

Learning content	Learning outcome	Learning process
Action	Change	Transformation
Values	Behaviours	Addressing challenges
Collaboration	Communication	Critical thinking

Source: UNESCO 2015

These core conceptual dimensions (as per Table 3.4) serve as the basis for defining Global Citizenship Education, learning objectives, and competencies, as well as priorities for assessing and evaluating learning. These core conceptual dimensions are based on, and include, aspects from all three domains of learning – cognitive, socio-emotional and behavioural – because global citizenship is not only about learning outcomes, but it is about competences and intelligences that go beyond the classroom activities.

The global citizen learner exercises the cognitive domain referring to thinking skills and the capacity of better understanding the world complexities, at the same time that this learner applies socio-emotional abilities related to values, social skills on how to live together with others respectfully and peacefully through the behavioural domain oriented to his conduct, performance, and engagement (UNESCO 2015).

These domains of learning derive from the pillars of learning from Delors (1996) in 'Learning: The Treasure Within':

a) Learning to know
b) to do
c) to be
d) to live together
e) to transform the world around

Considering the amplitude of components of global citizenship related to education, the implications for policy, curricula, teaching, and learning are enormous (Albala-Bertrand 1995; Banks 2004):

> Global citizenship education applies a lifelong learning perspective, believing that knowledge is important to all ages at all domains of application, from early childhood through all levels of education and into adulthood, involving formal and informal approaches, curricular and extracurricular interventions, and conventional and unconventional pathways to participation.
> (UNESCO 2014a, p. 46; 2015, p. 15)

> Global Citizenship Education takes 'a multifaceted approach, employing concepts and methodologies already applied in other areas, including human rights education, peace education, education for sustainable development and education for international understanding' and aims to advance their common objectives. Global Citizenship Education aims to be transformative, building the knowledge, skills, values and attitudes that learners need to be able to contribute to a more inclusive, just and peaceful world.
> (UNESCO 2014b, p. 14; 2015, p. 15)

While the global perspective is transformative and inclusive, it is worthwhile noting the potential of bioregions for addressing the Sustainable Development agenda. Bioregionalism or bioregional model is seen as an alternative strategy against the mechanistic thinking paradigm as it is defined by Aberley (1994) as a set of 'knowledge and practices intending to reconnect societies in a sustainable way with their local and regional natural matrix'. Its proposition is to have an integrative perspective of the local ecosystems, communities, and local economic activities when arguing that real sustainability depends on recognizing the elements and characteristics of the bioregion because 'every human community exists within a specific and unique bioregion consisting in the natural features that maintain life of that place' which has implications for the bioregional planning and management (Toledo 1999). Stakeholders oriented to a sustainable vision should have the global perspective but should also have the intelligence about the specific features and practices of different bioregions. This global-local perspective explains important convergences aligned to new models to address cross-border change and complexity as explained in Chapter 8 of this book.

3.5 Sustainable Development theory – cultural factor and education

SD and sustainability are dynamic concepts and processes dependent on the interconnected web of energy and life.

The origins of the Sustainable Development concept emerged as an attempt to balance economic, social, and environmental policies and law and to try to achieve a compromise for a more manageable resource exploitation according to conservationist approach aligned to 'Our common Future' ref, 'The Brundtland Report', and the 1992 UN Conference on Environment and Development (UNCED, Earth Summit). Through its principles, it has been recognized that humanity should live within the carrying capacity of the earth and manage natural resources to meet current and future needs of generations. The initial emphasis was related to the environmental debate and the balance among the pillars, but without defining at what extent this balance would be achieved and without touching the social component.

It was a time when the WCED (The World Commission on Environment and Development) formally recognized the interrelationships among the environmental, development, and energy crisis by stating:

> An environmental crisis, a development crisis, and energy crisis. They are all one. Ecology and economy are becoming ever more interwoven – locally, regionally, nationally, and globally – into a seamless net of causes and effects.
>
> (UNWCED 1987, p. 14)

It also recognized, in 1987, that social development priorities should consider social 'needs' and equitable 'opportunities for all':

> it is futile to attempt to deal with environmental problems without a broader perspective that encompasses the factors underlying world poverty and international inequality.
>
> (UNWCED 1987, p. 12)

Social issues and human rights were not fully considered, though it was recognized that the 'inability to promote the common interest in Sustainable Development is often a product of the relative neglect of economic and social justice within and among nations' (Segger & Khalfan 2006, p. 19).

On the other hand, sustainability is about recognizing the dynamic, cyclical, and interdependent nature of all the Earth's components, the interactions of humans with their habitats and the planet's bio-cycles. Sustainability is a dynamic state where the world becomes a place for everyone to pursue its right to live and thrive under a dynamic and durable equilibrium.

Sustainability can be understood as a system and a discipline. As a definition, it refers to the systems and processes that are able to operate and persist on their own over long periods of time. The term first appeared in the 1713 forestry book 'Sylvicultura Oeconomica' (Warde 2011, p. 153) where sustainable meant 'able to endure without failing for a long time'. The word 'sustainability' first appeared in 1972 to designate forms of human economic activity that does not lead to long-term environmental degradation and depletion of natural resources (The Oxford English Dictionary 2018).

From the definition comes the TBL or 'Triple Bottom Line' referring to Elkington's three pillars as planet, people, and profit or the three E's, ecology, economics, and equity (Elkington 2012, p. 55). The triple bottom line of sustainability recognizes the need to foster communities that are healthy, safe, and secure, with economic opportunity for everyone while keeping Earth's life support system in good shape (Cortese 2012, p. 12). Ecological integrity is vital for sustaining life according to limits or the carrying capacity. The economic pillar is achievable when resources are fairly distributed allowing individuals to meet their basic human needs. The third pillar meaning the social aspect, equity, or equality includes freedom – freedom from poverty, unhealthy living conditions, lack of employment, and poor education according to Section 3.3. It also includes intergenerational impact assessment not to affect negatively the future generations.

It seems that sustainability as a system has not considered along the time, the cultural factor, or the multicultural perspective. The social component has not been researched enough or explored since 1987, and it is an attempt to emphasize that no individual can realize its freedoms according to Nussbaum and Sen (1993) if his or her cultural values and practices are not considered (Robertson 2014, p. 7). Equity means to provide opportunities for all people to grow and flourish in their way.

If sustainability is theoretically compounded by three fundamental pillars, it is correct to conclude that it is an interdisciplinary or multidisciplinary field as it involves the environment, the economy, and the social dimensions of a specific matter. As in relation to the concept seen as a system, the interdisciplinarity of it is also a challenge in relation to its concept as a discipline. The main point of this challenge is to understand the complexity of the interactions of the three E's and understand that these interactions also embark other fundamental pillars like culture and politics. The trend so far has been to consider these three components in a compartmentalized way and to ignore the cultural and political fundamental pillars. It leads to the conclusion that there is a long path to follow in relation to social sciences research about sustainability and there are important changes to make into the concept through viewing it as a system, by linking it to Complexity Theory and to Capability Theory in the context of Higher Education.

Sustainable Development literacy builds upon a progression of environmental and ecological literacies (UNESCO 2015, p. 8). The basic premise of Sustainable Development Literacy is that human and natural systems are dynamically interdependent and cannot be considered in isolation in order to resolve critical issues through the application of process-based tools capable of managing unexpected change (Hutchins 1968, p. 1). As the geographical and cultural aspects are also relevant to define models of development, they seem to be equally important to define levels of Sustainable Development Literacy (Dale & Newman 2005, p. 352).

Terminology around sustainability is complex and culturally contested, with definitions varying according to context and perspective. Petrov et al. (2016, p. 165) argues that:

> Understanding the sustainability of Arctic social–ecological systems (SES), their ability to respond to external and internal pressures, and their adaptive capacities has become a key task for both researchers and Arctic stakeholders.
>
> (Petrov et al. 2016, p. 165)

This is the point when geographical knowledge defines the strategy and innovates the theory, the educational system, and the curriculum, because the 'cultural component' makes a huge difference in the Sustainable Development of the Arctic despite not being considered a pillar of the TBL. Cultural concerns are missing from the SD concept and from components of ESD.

3.5.1 Education for Sustainable Development (ESD)

Vare and Scott (2007) reflected on the need of thinking about different kinds of ESD, the relationship between educational outcomes (learning)

Table 3.5 Two sides of ESD.

ESD 1	ESD 2
• Promoting/facilitating **changes** in what we do	• Building capacity to **think critically** about [and beyond] what experts say and to test Sustainable Development ideas
• Promoting (informed, skilled) **behaviours** and **ways of thinking**, where the need for this is clearly identified and agreed	• Exploring the **contradictions** inherent in sustainable living
• Learning **for** Sustainable Development	• Learning **as** Sustainable Development

Source: Vare & Scott 2007, pp. 3, 4

Table 3.6 ESD1 and ESD2 – positives and negatives.

ESD 1	ESD 2
Positives	**Positives**
• Promotion of informed, skilled behaviours and ways of thinking	• Building capacity to think critically about what experts say and to test ideas
• Useful in the short-term where the need for this is clearly identified and agreed	• Exploring the dilemmas and contradictions inherent in sustainable living
	• A necessary complement to ESD 1, making it meaningful in a learning sense
Negatives	
• People rarely change their behaviour in response to a rational call to do so	
• Too much successful ESD 1 in isolation would reduce our capacity to manage change ourselves and therefore make us less sustainable	

Source: Vare & Scott 2007, pp. 3, 4

and social change (behaviour change). In other words, the authors felt the need to define grades how to assess levels of ESD in terms of the education *praxis*, based on specific elements in bold.

Initially, they proposed the ideas of ESD 1 and ESD 2 disposed as a complementary framework and adapted to a consolidated version in Tables 3.5 and 3.6, which show a shorter version to summarize and systematize the two types of ESD.

In recent reflection about SD, learning, and change, Scott and Gough (2003, pp. 113, 116) identified three types of approaches that shape the framework summarized as per Table 3.7.

Table 3.7 Three types of ESD based on Scott and Gough (2003, pp. 113, 116)

Type 1 – learning about SD	Type 2 – learning for SD	Type 3 – learning as SD
• Assume that the problems humanity faces are essentially **environmental**	• Assume that our fundamental problems are social and/or political, and that these problems produce **environmental symptoms**	• Assume that what is (and can be) known in the present is **not adequate**
• Can be understood through science and resolved by appropriate environmental and/or social actions and **technologies**	• Such fundamental problems can be understood by means of anything from **social-scientific analysis** to an appeal to indigenous knowledge	• Desired 'end-states' cannot be specified
• It is assumed that learning leads to change once facts have been established and **people are told what they are**	• The solution in each case is to bring about **social change**	• This means that any learning must be **open-ended**
	• Learning is a **tool to facilitate choice** between alternative futures which can be specified on the basis of what is known in the present	• Essential if the uncertainties and complexities inherent in how we live now are to lead to **reflective social learning** about how we might live in the future

Tables 3.5, 3.6, and 3.7 are useful to understand the opportunities and challenges of the possible types of ESD, based on specific features in bold; however, these seem oversimplified in the face of the elements and contexts under research in this book what makes the case for a reassessment of the ESD components towards a more comprehensive model. This is an important contribution of this research as the complexity of the geographical Arctic landscape, in its several components – economic, social, environmental, **and cultural** – defines specific approaches, pedagogies, educational experiences, and curriculum in relation to the local reality of SD.

HE has educated the politicians, managers, teachers, scientists, and engineers who have taken societies to our current unsustainable position, and it is the education of future groups that will enable people to step up to new levels of sustainability. For this reason, curriculum change towards sustainability-related topics is the critical next stage (Sheppard 2015, p. 2). This argument is reinforced by the fact that according to Sheppard (2015, p. 4) there is a misalignment between ESD and HE.

These gaps need to be addressed as government departments, businesses, and communities are often not prepared for complexity-oriented decisions involving risks derived from extractive industry growth mainly in a sensitive area like the Arctic Circle. Legal and political frameworks reflect a scenario requiring improvement in terms of structure, guidelines, standards, and legislation. Specific skills and expertise are being required to negotiate and implement policies, to conduct the socio-environmental assessments, to formulate new legislation, and to consolidate an open dialogue between stakeholders. Arctic geo-capabilities are an essential component to promote Arctic Sustainable Development. Because they are especially important for communities, frequently expected to engage on complex issues, without having had the chance to discuss these within the community, which would help people develop their own opinions and expectations on particular projects and fully understand the implications of decision-making.

This chapter presented the theoretical framework underpinning the research project. The first aspect identified was the complexity of Arctic systems and the dynamic nature of change experienced by bio-physical and socio-cultural systems in the Arctic and the importance of capturing this perspective to expand the local understanding through the development of innovate capabilities.

The relevance of using Complexity Theory is to usefully tie together the curriculum design process in the context of Arctic ecosystems, capabilities, and ESD/EL by looking at the process of curriculum design in specific Arctic places on specific courses as a lens for understanding the more complex system of HE, Arctic change, and Arctic energy production and literacy.

The second aspect highlighted was the importance of a holistic and geographical approach into the local knowledge under a systematic way of analyzing change considering holism as a process of creative synthesis generating powerful knowledge.

The third aspect assumed that development should be seen as a way to expand what people are able to do and to be or their real freedoms (Capability Theory). This is a human-centred view of development where a healthy economy is one that enables people to enjoy a long and healthy life, a good education, a meaningful job, physical safety, democratic debate, and so on. This is what configurates elements of citizenship. For this reason, Global Citizenship Education (the fourth relevant theory) seems to amalgamate all these important characteristics providing a balance to TBL (the fifth relevant aspect) and supplying other components lacking in the current ESD models. Finally, constructivism and interpretivism are relevant theoretical assumptions of this book, as different cultural groups and different styles of education are under consideration.

References

Aberley, D 1994, *Futures by design: the practice of ecological planning*, New Society Publishers, Gabriola Island.

Albala-Bertrand, L 1995, *What education for what citizenship? First lessons from the research phase*, Educational Innovation and Information, Geneva, UNESCO IBE, No. 82.

Alkire, S & Deneulin, S 2009, 'The normative framework for development', in S Deneulin & WL Shahani (eds.), *An introduction to the human development and capability approach*, Earthscan, London, pp. 28, 35.

Arruda, GM & Krutkowski, S 2016, 'Arctic governance, indigenous knowledge, science and technology in times of climate change: self-realization, recognition, representativeness', *Journal of Enterprising Communities: People and Places in the Global Economy*, vol. 11, no. 4, pp. 514–528.

Arruda, GM & Krutkowski, S 2017, 'Social impacts of climate change and resource development in the Arctic: implications for Arctic governance', *Journal of Enterprising Communities: People and Places in the Global Economy*, vol. 11, no. 2, pp. 277–288.

Banks, J 2004, *Diversity and citizenship education: global perspectives*, John Wiley & Sons, San Francisco, CA.

Board of Education 1931, *Primary education*, Hadow Report on Primary Education, HMSO, London, p. 75.

Boni, A & Walker, M 2013, *Human development and capabilities: re-imagining the university of the twenty-first century*, Routledge, New York, p. 20.

Boulton, J, Bowman, C & Allen, P 2015, *Embracing complexity: strategic perspectives for an age of turbulence*, Oxford University Press, Oxford, UK, pp. 1, 10, 11, 20, 29, 35, 63, 66, 117.

Cortese, A 2012, 'Foreword', in H Hemderson (ed.), *Becoming a green professional: a guide to careers in sustainable architecture, development and more*, John Willey, New York, pp. 11–13.

Dale, A & Newman, L 2005, 'Sustainable development, education and literacy', *International Journal of Sustainability in Higher Education*, vol. 6, no. 4, pp. 351–362.

Davis, B 2008, 'Complexity and education: vital simultaneities', in M Mason (ed.), *Complexity theory and the philosophy of education*, Wiley-Balckwell, Oxford, p. 47.

Delors, J, et al. 1996, *L'éducation: Un trésor est caché dedans*. 1ére Edition. Éditions Odile Jacob, Paris.

Delors, J, et al. 1996, 'Learning: the treasure within'. Report to UNESCO of the international commission on education for the twenty-first century. UNESCO, Paris, p. 37

DeWaters, JE & Powers, SE 2011, 'Energy literacy of secondary students in New York State (USA): a measure of knowledge, affect, and behavior', *Energy Policy*, vol. 39, pp. 1699–1710.

Dill, J 2013, *The longings and limits of global citizenship education*, Routledge, New York.

Doll, WA 1993, *A postmodern perspective on curriculum*, Teachers College Press, New York and London, p. 11.

Donert, K 2015, 'GeoCapabilities: empowering teachers and subject leaders', The Innovative Pedagogies Series, York, Higher Education Academy, p. 2, viewed 12 December 2016, www.heacademy.ac.uk/geocapabilities-empowering-teachers-subject-leaders.

Elkind, D 2004, 'The problem with constructivism', *The Educational Forum*, vol. 68, pp. 306–312.
Elkington, J 2012, *The Zeronauts: breaking the sustainability barrier*, Routledge, New York, p. 55.
Fukuda-Parr, S & Kumar, SA 2003, *Readings in human development*, Oxford University Press, Delhi, p. 7.
Fullan, M 1989, 'Managing curriculum change', in M Preedy (ed.), *Approaches to curriculum management*, Open University Press, Milton Keynes, p. 144.
Goleman, D 2009, *Ecological intelligence. How knowing the hidden impacts of what we buy can change everything*, Broadway Books, New York.
Haigh, M 2014, 'From internationalisation to education for global citizenship: a multi-layered history', *Higher Education Quarterly*, vol. 68, no. 1, pp. 6–27.
Haq, M 2004, 'The human development paradigm', in S Fukuda-Parr & AK Shiva Kumar (eds.), *Readings in human development*, 2nd edn, Oxford University Press, New Delhi, pp. 17, 19.
Hein, G 1991, 'Constructivist learning theory', viewed 20 June 2018, www.exploratorium.edu/ifi/resources/constructivistlearning.html.
Hinchliffe, G & Terzi, L 2009, 'Introduction to the special issue "Capabilities in education"', *Studies in Philosophy and Education*, vol. 28, pp. 387–390.
Hutchins, R 1968, *The learning society*, University of Chicago Press, Chicago, IL, p. 1.
Kelly, AV 2009, *The curriculum. Theory and practice*, 6th edn, Sage, London, pp. 88, 99.
Lambert, D 2014, 'Curriculum thinking, 'capabilities' and the place of geographical knowledge in schools', *Prace Komisji Edukacji Geograficznej*, vol. 3, pp. 13–30.
Lambert, D, Solem, M & Tani, S 2015, 'Achieving human potential through geography education, a capabilities approach to curriculum making in schools', *Annals of the Association of American Geographers*, p. 217, doi: 10.1080/00045608.2015.1022128.
Marsh, CJ 2009, *Key concepts for understanding curriculum*, 4th edn, Routledge, Oxford, p. 7.
Mason, M 2008, *Complexity theory and the philosophy of education*, Wiley-Blackwell, Oxford, pp. 12, 33, 45, 216.
Maude, A 2015, '"What is powerful knowledge, and can it be found in the Australian geography curriculum?" Flinders University, Adelaide', *South Australia Geographical Education*, vol. 28, pp. 18–26, 20.
Mezirow, J 2009, 'An overview of transformative learning', in K Illeris (ed.), *Contemporary theories of learning*, Routledge, Abingdon, pp. 90–105.
Morrison, K 2008, 'Educational philosophy and the challenge of complexity theory', in M Mason (ed.), *Complexity theory and the philosophy of education*, Wiley-Balckwell, Oxford, pp. 16, 29, 45.
Moses, J (2004) 'Complexity and flexibility', paper presented at the 2004 Engineering Systems Symposium, MIT, Cambridge, p. 2.
Nussbaum, M 2000, *Women and human development: the capabilities approach*, Harvard University Press, Cambridge, MA, pp. 70–77, 72, 73, 74.
Nussbaum, M 2006, *Frontiers of justice: disability, nationality, species membership*, Harvard University Press, Cambridge, MA, pp. 78–81.
Nussbaum, M 2011, *Creating capabilities: the human development approach*, Harvard University Press, Cambridge, MA, pp. 33–34.
Nussbaum, M & Sen, A 1993, *The quality of life*, Clarendon Press Oxford University Press, Oxford, England and New York, p. 20.

Osberg, D & Biesta, G 2007, 'Beyond presence: epistemological and pedagogical implications of "strong" emergence', *Interchange*, vol. 38, no. 1, pp. 31–55.

Osberg, D, Biesta, G & Cilliers, P 2008, 'From representation to emergence: complexity's challenge to the epistemology of schooling', *Educational Philosophy and Theory*, vol. 40, no. 1, pp. 213–227.

Petrov, A, BurnSilver, S, Chapin, F, Fondahl, G, Graybill, J, Keil, K, Nilsson, A, Riedlsperger, R & Schweitzer, P 2016, 'Arctic sustainability research: toward a new agenda', *Polar Geography*, vol. 39, no. 3, pp. 165–178.

Prigogine, I 1980, *From Being to Becoming*. W.H. Freeman and Company, New York.

Pittman, J 2017, 'Exploring global citizenship theories to advance educational, social, economic and environmental justice'. *Journal of Tourism & Hospitality*, vol. 6, p. 326.

Roberts, A, Cherry, J, Scott, E, Hinzman, L & Walsh, J 2010, *Proposed development of a Community Arctic System Model for understanding environmental complexity*, State of the Arctic: Approaches to Integrated Studies of the Arctic System, International Arctic Research Center, Miami, FL, March 16, p. 3.

Robertson, M 2014, *Sustainability principles and practice*, 1st edn, Routledge, Oxford, p. 7.

Robeyns, I 2005, 'The capability approach: a theoretical survey', *Journal of Human Development*, vol. 6, no. 93–114, pp. 94, 96.

Robeyns, I 2016, 'The capability approach', The Stanford Encyclopedia of Philosophy (Winter 2016 Edition), Edward N. Zalta (ed.), p. 3, viewed 13 January 2016, https://plato.stanford.edu/archives/win2016/entries/capability-approach/.

Ryan, A 2011, 'ESD and holistic curriculum change', pp. 3, 5, viewed 12 December 2016, www.heacademy.ac.uk/resources/detail/sustainability/esd_ryan_holistic.

Santonus, M 1998, 'Simple, yet complex', p. 3, viewed 7 December 2018, www.cio.com/archive/enterprise/041598_qanda_content.html.

Scott, WAH & Gough, SR 2003, *Sustainable development and learning: framing the issues*, RoutledgeFalmer, London, pp. 113, 116.

Segger, MC & Khalfan, A 2006, *Sustainable development law. Principles, practices & prospects*, OUP, Oxford, p. 19.

Sen, A 1992, *Inequality re-examined*, Claredon Press, Oxford, pp. 5, 39, 40, 48.

Sen, A 1999, *Development as freedom*, Oxford University Press, Oxford, pp. 19, 75, 87.

Sheppard, K 2015, *Higher education for sustainable development*, Palgrave Macmillan, Basingstoke, pp. 2, 4.

Smuts, J 1999, *Holism and evolution*, Sierra Sunrise Publishing, Sherman Oaks, CA, p. 87.

Solem, M, Lambert, D & Tani, S 2013, 'GeoCapabilities: toward an international framework for researching the purposes and values of geography education', *Review of International Geographical Education Online*, vol. 3, no. 3, pp. 204–219.

Spencer, H 1864, *Principles of biology. v. 1*, September 1864, D. Appleton and Company, New York, p. 444.

St. Clair, R 2003, 'Words for the world: creating critical environmental literacy for adults', *New Directions for Adult and Continuing Education*, vol. 2003, no. 99, pp. 69–78.

The Oxford English Dictionary 2018, www.oxforddictionaries.com/oed

Tilbury, D 2011, *Higher education in the world 4. Higher education's commitment to sustainability*, Barcelona Palgrave and Global Universities Network for Innovation (GUNI), p. 3.

Toledo, V 1999, *Consensos Naturo-Sociales: Una Evaluación de las Nuevas Construcciones del Territorio y de las Regiones*, Comité Técnico Inter-agencial del Foro de Ministros de Medio Ambiente de América Latina y el Caribe, doc, Mimeo, julio, p. 1999.

UNDP (United Nations Development Program) 1990, *Human development report*, OUP, New York and Oxford, p. 9.

UNESCO 2014a, 'Education strategy 2014–2021', UNESCO, Paris, p. 46, viewed 2 December 2018, www.natcom.gov.jo/sites/default/files/231288e.pdf.

UNESCO 2014b, *Global citizenship education. Preparing learners for the challenges of the 21st century*, UNESCO, Paris, p. 14.

UNESCO 2015, *Global citizenship education: topics and learning objectives*, UNESCO, Paris, pp. 8, 15, viewed 26 November 2018, https://unesdoc.unesco.org/ark:/48223/pf0000232993.

UNESCO 2016, 'Schools in action. Global citizens for sustainable development. A guide for students', UNESCO, Paris, p. 6, viewed 25 November 2018, www.unesco-sole.si/doc/teme/sdg-guide-for-teachers.pdf.

UNWCED 1987, 'Report of the World Commission on Environment and Development', General Assembly Resolution 42/187, 11 December 1987, pp. 10, 12, 14, viewed 30 March 2018, www.un.org/documents/ga/res/42/ares42-187.htm.

van der Horst, D, Harrison, C, Staddon, S & Wood, G 2016, 'Improving energy literacy through student-led fieldwork – at home', *Journal of Geography in Higher Education*, vol. 40, no. 1, pp. 67–76.

Vare, P & Scott, WAH 2007, 'Learning for a change: exploring the relationship between education and sustainable development', *Journal of Education for Sustainable Development*, vol. 1, no. 2, pp. 3, 4.

von Bertalanffy, L 1969, *General system theory*, George Brazillier, New York, p. 12.

Vygotsky, LS 1978, *Mind in society: development of higher psychological processes*, Harvard University Press, Cambridge, MA, pp. 57, 59.

Wadley, D 2008, 'The garden of peace', *Annals of the Association of American Geographers*, vol. 98, no. 3, p. 650.

Waldrop, MM 1993, *Complexity: the emerging science at the edge of order and chaos*, Penguin, Harmondsworth, p. 88.

Walkington, H, Dyer, S, Solem, M, Haigh, M & Waddington, S 2018, 'A capabilities approach to higher education: geocapabilities and implications for geography curricula', *Journal of Geography in Higher Education*, vol. 42, no. 1, pp. 7–24.

Warde, P 2011, 'The invention of sustainability', *Modern Intellectual History*, vol. 8, pp. 153–170.

Wheatley, M 1999, *Leadership and the new science: discovering order in a chaotic world*, 2nd edn, Berret-Koehler Publishers, San Franscisco, CA, p. 10.

Young, M 2008, 'From constructivism to realism in the sociology of the curriculum', *Review of Research in Education*, vol. 32, no. 1–32, p. 14.

Young, M 2011, 'The future of education in a knowledge society: the radical case for a subject-based curriculum', *Journal of the Pacific Circle Consortium for Education*, vol. 22, no. 1, pp. 21–32.

Young, M 2013, 'Powerful knowledge: an analytically useful concept or just a "sexy sounding term?" A response to John Beck's "Powerful knowledge, esoteric knowledge, curriculum knowledge"', *Cambridge Journal of Education*, vol. 43, pp. 195–198, 196.

Young, M 2014, 'Powerful knowledge as a curriculum principle', in M Young, D Lambert, C Roberts & M Roberts (eds.), *Knowledge and the future school: curriculum and social justice*, Bloomsbury Academic, London, pp. 65–88, 74.

Youngblood, M 1997, *Life at the edge of chaos*, Perceval Publishing, Dallas, TX, p. 27.

4 The nature of the Arctic energy curriculum

4.1 Introduction

The findings from academic staff members' interviews are presented in this chapter helping to uncover important components concerning the objectives of each course and the nature of their specific curricula according to academic staff's teaching experience and practices in the case study locations. The importance of codes was determined based on the interpretation of components shared by the greatest number of participants.

Through this chapter it is possible to recognise that different people have different personal, social and cultural constructions of reality. In this context, interpretivism offered a useful methodological approach in order to analyse and interpret both indigenous and western perspectives regarding the data. The adoption of an interpretivist method for the analysis allowed to look across the data from multiple institutions and draw out insights. In this sense, interpretivism was of assistance to interpret data. Social constructivism often combined with interpretivism is typically seen as an approach to qualitative research (Bruner 1996; Lave & Wenger 1991; Mertens 1998; Rogoff 1990). As well as essentially valuing the diverse opinions and belief systems of individuals, illuminative evaluation emphasizes the importance of context in gaining an understanding of the education programme concerned. This means that a central part of the investigative process involves the researcher in actively discovering the social, institutional, political and cultural context of the education programme in all its complexities (Parlett 1981).

Table 4.1 shows the lecturers' code names, their teaching locations in terms of countries and universities where the interviews took place. The table also shows the 'research population type' in terms of the public involved in the teaching (program) and the public reached by the program as non-indigenous, indigenous or mixed public meaning that the students are mixed (non-indigenous and indigenous studying the same course). This characteristic was important to understand the cultural driver or educational environment impacting the interview responses.

The findings being presented describe the content of the interviews carried out with lecturers. The counterpoints, using the data collected from

Table 4.1 A summary of the lecturers' code names, educational institutions, countries and education environment

Lecturer code name	University	Country	Institution Population type
Nils (non-indigenous)	Borealis University	Norway	non-indigenous
Arne (non-indigenous)	Borealis University	Norway	non-indigenous
Randi (non-indigenous)	Borealis University	Norway	non-indigenous
Ove (non-indigenous)	Borealis University	Norway	non-indigenous
Alf (non-indigenous)	Borealis University	Norway	non-indigenous
Kristine (indigenous)	Glacial University	Norway	Indigenous
Eva (indigenous)	Glacial University	Norway	Indigenous
Ivar (indigenous)	Glacial University	Norway	Indigenous
Line (indigenous)	Glacial University	Norway	Indigenous
Grete (indigenous)	Glacial University	Norway	Indigenous
Felix (non-indigenous)	Juniper University	Canada	Mixed
Zach (non-indigenous)	Juniper University	Canada	Mixed
Emma (non-indigenous)	Juniper University	Canada	Mixed
Rayan (non-indigenous)	Juniper University	Canada	Mixed
Noah (non-indigenous)	Ice University	Alaska	Mixed
Olivia (indigenous)	Ice University	Alaska	Mixed
Thomas (non-indigenous)	Ice University	Alaska	Mixed
Audrey (non-indigenous)	Ice University	Alaska	Mixed

the interviews with students in the informed locations, are made in Chapter 8 and the data will serve to answer the research questions extracted from the research objectives as per presented in Chapter 1. This chapter serves as a guideline to orient the presentation of findings regarding the nature of the energy curriculum and its relationship with the human and geographical components. In other words, this chapter provides subsidies to understand at what extent the energy curriculum is connected to the human dimension and considers societal aspects. A second fundamental aspect of this chapter was to evaluate embedded levels of energy literacy. Levels of energy literacy and levels of adherence to ESD will be assessed in Chapter 7 by comparing interview data with data collected from courses documents.

4.2 The nature of the energy curriculum

4.2.1 Key objectives of the course as per lecturer's perspective- indigenous vs. non-indigenous education

When outlining the nature of the energy curriculum, the first central element is the key objectives of the courses. The nature and characteristics of

110 *The nature of the Arctic energy curriculum*

the curriculum are highly determined by the objectives it aims to accomplish. Curriculum is usually viewed as an instrument of change and this section outlines the characteristics extracted from the interviews that define the nature of the energy curriculum in sustainability (or human geography)-related programmes in each of the studied locations as per the coding process. The codes represent the main themes extracted from the interviews with academic staff. Table 4.2 shows the codes representing the influencing

Table 4.2 The most influencing factors of course objectives in the case study universities. The objectives are sequenced so that those mentioned by the largest proportion of participants appear first

Code	Definition
1. To understand the concept of SD	• To study what Sustainable Development means to the indigenous people – Sustainable Development means the continuation of indigenous culture and livelihoods in an economically sustainable way. • A standard notion that means to ensure that citizens of a particular area understand their development and that the use of resources need to be balanced between the needs of the current generation and the future generations. • A term meaning how humans can have a low impact on Earth and living within the limits of ecosystems not using up resources such as carbon-based fuels and looking to Renewable Energy and to avoid farming practices that deplete the soil and cause trouble for future population. • To learn the formal and international concept of SD according to international treaties and formal definitions, principles and characteristics.
2. To make students aware of SD and energy issues	• Our developmental and energy systems are based on our economic activities and cultural backgrounds.
3. To understand the interactions among natural resources, energy, environment and society	• To be able to produce environmental assessments on the interactions/impacts of energy, natural resources on the local environment.
4. To make students aware of natural resources management	• How to manage resources in a sustainable way. • To provide knowledge about the impacts of extractive activities on land that sustains people's livelihoods.
5. Intergenerational impacts	• How our use of energy/natural resources impacts our life-supporting systems and future generations.
6. Critical thinking	• To form people who are able to look at academic literature and able to make a reasonable research-based argument.

factors identified in the interviews with staff at the four case study universities. The codes are presented in order of importance determined by being most widely shared across the sample and are explained in more detail in the following sections of this chapter.

By examining the findings acquired from the interviews with lecturers, the codes, presented in Table 4.2, were identified.

4.2.1.1 Code 1 – to understand the concept of Sustainable Development (SD)

The understanding of SD concept seems to be an important objective for the lecturers of the four universities under study. At the same time the concept is relevant for their courses it is also, in their views, a difficult term to define and to make students understand. This is due to the fact that different people see this term in different ways according to their own economic and cultural systems that provide different perspectives.

The consequence of these different perspectives regarding what is sustainable implies that SD means different things to different people, but all the lecturers pointed out the importance of providing the students with a clear understanding of the concept of SD as a primary objective of their courses.

For the Glacial University lecturers, there is a particular understanding of SD, not exactly like in the westernized education but within the indigenous environmental knowledge and based on indigenous natural resources management and the observation of the climatic changes currently in place in the Arctic environment. To them, Sustainable Development means the continuation of their indigenous culture and livelihoods in an economically sustainable way. The course focussed on what Sustainable Development means to the indigenous people. These lecturers also called attention to a gap in the concept's interpretation.

GRETE (INDIGENOUS): Sustainable Development is a difficult term to understand because we see it in a different way. We have our perspective about what is sustainable. We have a particular understanding here. It is not exactly like in the westernized education, but we see it within the Indigenous environmental knowledge and the changes in our environment. I think the indigenous approach and methodology to education and sustainability aims at making people aware of this gap in the understanding.

KRISTINE (INDIGENOUS): In our community, I think that Sustainable Development means that we will leave and consume only what we need, and not excessively, and leave the place to our children like we got it from our fathers.

RAYAN (NON-INDIGENOUS): (…) the development in the way that protects the ability of future generations to survive and thrive and to develop further without bearing significant consequences of our actions today.

These lecturers were clear about the fact that not everybody understands the meaning of Sustainable Development because the concept lost the meaning, being just a political phrase according to one lecturer (Ivar):

IVAR (INDIGENOUS): Sustainable Development is just a political concept.

He argues that SD relates to a longer period of time that is not comprehensible to the human-being's mind. The individual does not understand 500 years, or what may happen in 100 years. Human beings do not comprehend the impact of actions and decisions taken presently nor the future effects they may cause by producing energy or using natural resources. Ivar believes that the term 'sustainable' would mean that we can use the resources not only today, not only tomorrow, not the day after but for over a very long period of time, that he cannot define or quantify but over a long time what could mean 50.000 years from now, without destroying the resources of the environment we depend on. The mind-set does not reach this timeframe or the SD time dimension because the human being's mind is unable to comprehend the intergenerational impacts of his/her actions:

IVAR (INDIGENOUS): the term Sustainable Development comes from Our Common Future, it was first used in Our common future in the 1987, Brundtland commission, but now the concept is a bit difficult because you can use it for everything, to be politically right you need to be sustainable, but Sustainable Development, as a concept, lost the meaning, it is just a political phrase many use, because it is expected that, as a politician, you shall mention Sustainable Development, but I am not sure if everybody understands what Sustainable Development is. Sustainable Development is not only for the 4-year-time of an election. Natural science has a longer timeframe.

Like Ivar (indigenous), Nils (non-indigenous) argues that politicians have used the term in relation to short-term political campaigns and mandates (4 years) and in debates about oil and gas exploration in Arctic locations, but the question is if fossil resources would be considered sustainable, not only because they represent finite resources but also because of the long-term impacts they cause. Nils also emphasizes the idea that the concept means different things to different people by stating that:

NILS (INDIGENOUS): Sustainable Development is different, Sustainable Development for politicians, for geographers, sociologists, especially for ecologists; the mentality about this phenomenon 'Sustainable Development' is different. In my idea Sustainable Development is some method to adapt our livelihood to the capacity of natural resources.

In the same vein and emphasizing the intergenerational component, Thomas (non-indigenous) argues that 'for him, Sustainable Development means forever':

THOMAS (NON-INDIGENOUS): Sustainable means forever; something that can be done forever; development forever; systems that rejuvenate; renewable resources. You cannot be using non-renewable resources and have sustainable development.

The intergenerational timeframe of impacts and the ability of future generations to survive and thrive lead us to another important element that is the capacity to rejuvenate and be renewed. Technically, this relates to the ecosystems' resilience capacity dependent on a low footprint on Earth and on the ability to live within the limits of ecosystems:

AUDREY (NON-INDIGENOUS): It is a term meaning how humans can have a low impact on Earth and living within the limits of ecosystems not using up resources such as carbon-based fuels and looking to Renewable Energy and avoiding farming practices that deplete the soil and cause trouble for future population.

In Canada the interviews indicated that a more westernized and standardized concept of SD is accepted and adopted in teaching. The lecturers referred to the main components of the Brundland Report and some of them briefly paraphrased the main elements of the Brundtland concept of SD related to patterns of production and consumption, the levels of waste generated by the present patterns of production and consumption, the importance of a balanced use of natural resources as well as the intergenerational aspect of the Brundtland Report concept of SD. Examples are:

EMMA (NON-INDIGENOUS): I think this is a standard notion and it is to ensure that citizens of a particular area understand that development; particularly, the use of resources need to be balanced between the needs of the current generation and the future generations.
FELIX (NON-INDIGENOUS): I have thought about it very much and it means to me more modest level of consumption and less waste in production and consumption.
RAYAN (NON-INDIGENOUS): the development in the way that protects the ability of future generations to survive and thrive and to develop further without bearing significant consequences of our actions today.

On the other hand, there was a particular answer that demonstrated vagueness in relation to the understanding of the concept of SD.

ZACH (NON-INDIGENOUS): This is a good question. I would not think of a definition, it depends on who's trying to save what. The traditional definition, the Brundtland definition is a widely accepted one but I don't think I have strong thoughts on it.

In summary, while there is an educational concern to communicate the concept of SD in a clear way to students as a shared objective of the courses, lecturers have very different understandings about the concept according to their economic and cultural systems and contexts.

4.2.1.2 Code 2 – to make students aware of SD and energy issues: ways of doing it in education

This code basically emphasises another common objective among the courses delivered in different locations. Awareness is seen as the importance of transmitting to the students the notions regarding SD, the interrelations to energy and the importance of conveying the notion and the connections, in an efficient way.

Most lecturers interviewed identified two dimensions of this awareness: awareness related to the environment and its changes and awareness related to society. They consider this capacity of making students aware of SD and energy issues as a very important objective of the courses in place justified by adaptation and knowledge dissemination needs. Examples of lecturers' words in this regard are:

GRETE (INDIGENOUS): I consider this discussion very important at the moment as we are experiencing effects of climate change and these effects are very visible in our area and in our lives. The youth need to be educated about these changes and how to adapt to them. Then they are disseminators of their knowledge to their communities and they can help in the process.
EVA (INDIGENOUS): The importance of awareness of SD has two dimensions: society as a whole and local communities. Education for Sustainable Development means learning about the environmental impacts of human activity. For me, Education for Sustainable Development means also to work on empowering indigenous youth to bring their livelihoods and culture to the future.

According to the main point presented by the lecturers, human beings think based on a very short period. The short term thinking of mankind does not consider that the benefits of today can represent a serious damage for the next generation. The relevance of creating awareness about the intergenerational socio-environmental impacts and economic effects of those impacts seem to guide lecturers to reflect on collective responsibility in relation to development in society, alternative learnings about how to live in society, the student's role in Sustainable Development and the students' role towards a sustainable world.

On one side, the lecturers recognise the importance of being aware of SD related to energy issues, on the other they diverge in relation to the way of doing it. Some lecturers pointed out that SD should be the mission of geographers. Basically, their position indicated that Higher Education should be more focused on SD and awareness could be inspired and raised by a more institutionalized way of approaching the subject via ESD or energy literacy (EL) – as the main component of SDL. These are common patterns especially related to participants from Norway and Alaska on how SD should be more emphasized in education through formal mechanisms of literacy that have a significant impact not only on the individual but on the collective spectrum (individual and communities' awareness).

The justification for this emphasis relates to the fact that the awareness of environmental and societal issues is a determinant on the student's attitude in face of his/her awareness, the time of action in relation to the awareness and the quality of the decisions students make when aware of specific issues. In other words, lecturers described the relevance of creating and enhancing awareness through a more systematic educational tool like basic literacy regarding Sustainable Development/energy in order to orient student's decision-making in the future, how students make educated decisions on crucial aspects of their lives and the type/quality of choices people make when aware of the trade-offs, their impact and on how to reduce their own footprint. Despite having reflected about how to raise awareness and agreeing with this position, it does not mean that these mechanisms are in place in all locations studied. Noah, stressed the relevance of awareness through literacy for students making good and educated decisions as follows:

NOAH (NON-INDIGENOUS): I think it has a huge relevance not just for undergraduate students. I think that everybody should have basic literacy regarding Sustainable Development because if people do have energy literacy, they will be able to make educated decisions and make simple actions in their lives that can have a very significant impact. So, I think the relevance is huge not only for undergraduate students but for everybody.

RANDI (NON-INDIGENOUS): I think it is important, but we have left it to those that work with technology and we are not taking enough on board within the social sciences – it needs to be institutionalized and it needs to be done systematically.

Other lecturers that advocate the importance of making students aware of SD and energy issues, raised questions on how to incorporate SD/energy into formal education and how to incorporate these elements into the curriculum. For them, sustainability does not happen in an education setting, but in the real world. They justify this position saying that Sustainable Development is a concept that varies according to the context and culture. Sustainability (as a function of SD) for them is a cultural value or, in other words, SD is defined by culture and it is hard to teach values to people. Kristine is a good example of this way of thinking when saying:

KRISTINE (INDIGENOUS): I think Sustainability is a value, a cultural value and it is defined by the culture, the community you grow up and development of sustainability is hard to teach, because it involves teaching about values; how do you teach values? by doing or by showing good examples. Do you teach values by talking about it? How do you teach that? What is the way to incorporate this into formal education?

Kristine (indigenous) suggests, with her questions, that it is not possible to inculcate values that are transmitted by generations and family tradition, through teaching. She suggests that values can only be taught by practice, not by talking about them. According to her mind-set values that are only part of an educational rhetoric are not assimilated by the students and these values are difficult to incorporate in a lecture-based westernized style of education.

The reason for Kristine's questions points to the fact that indigenous values are different in relation to Sustainable Development and indigenous lecturers find it very difficult in a western education system to incorporate more than one perspective in formal education. Glacial University lecturers believed that there is the need for students to be taught under different viewpoints as the present western education system considers just one viewpoint. This presents a crucial question on how to make students aware of SD and energy issues if a fundamental belief, such as the sustainability of a traditional and transgenerational economic activity is contested in class. Indigenous people believe that economic activities like fishing, hunting and reindeer husbandry are sustainable activities because for several generations their ancestors have lived off the sea and the land; they believe that whales, seals, reindeers don't destroy the environment, and they are considered a renewable source, if the ecosystems' limits are maintained, as peoples have lived for millennia in the territories developing these types of activities that provide food, jobs, income, livelihood. The lecturers believe that energy extraction/mining depletes the land and the environment. These lecturers believe indigenous economic activities are sustainable, for specific cultural reasons, but in western education lecturers teach differently in class and data can be found saying the contrary in all sorts of publications. An illustration of this clash in cultural values impacting on teaching is the fact that in some northern universities operating under a more westernized style of teaching, some indigenous students studying courses related to environmental and natural resources had experiences of lecturers stating that indigenous economic activities were considered harmful to the environment. The indigenous view is that they have practiced these activities for millennia and their peoples have survived more than 5000 years by practising their activities respecting the intergenerational sustainability and ecosystem capacity. This creates a confusion and a contradiction in indigenous students' learning experience.

4.2.1.3 Code 3 – to understand the interactions among natural resources, energy and environment

Another common aspect that was reported by the interviewees was the importance of providing students with an understanding of the interactions among natural resources, energy and the environment. The dominant point is related to the impacts of these interactions in terms of variables of space (Arctic geographical area) and time. There are, however, different justifications that motivate this objective in the studied courses' objectives.

For some lecturers this understanding provides that SD is translated into a more local actionable scheme. Since the students are provided with the right tools to understand the levels and consequences of the interactions among natural resources, energy and environment it will ensure their social impact and public participation in making decisions on these matters and within actual projects. One important reason for this approach is climate change. According to Nils:

NILS (NON-INDIGENOUS): Sustainable Development is much linked to ecology, in geography, but ecology is just as studying and thinking in small scale; not enough for solving big problems like climate change.

Nils (non-indigenous) believes that more discussion about this is required. It does not seem that the curriculum covers the scale of these contemporary issues – the curriculum does not cover these big scale problems as they deserve. In general, lecturers from the three studied locations agree that it is important to develop more comprehensive and engaging ways to understand the impacts from the interactions of natural resources, energy and environment under the auspices of Climate Change.

Aligned to this argument, lecturers also reported the importance of providing discussion on patterns of production and consumption and the benefits of local consumption versus importing goods from long distances in order to reduce Greenhouse Gases emissions and, consequently, mitigate the negative effects of climate change locally and globally. The importance of sustainable food production under sustainable farming practices to reduce emissions are some of Felix's thoughts, as an example:

FELIX (NON-INDIGENOUS): I have thought about it very much and it means to me more modest level of consumption and less waste in production and consumption.

Another emphasized point was the importance of providing a clear understanding about energy production, consumption and use. Some lecturers mentioned that they approach this topic by enhancing understanding on how to assess the positive and negative impacts of energy production and use in the long term and the mechanisms of reducing energy consumption and enhancing energy efficiency.

For some interviewees the importance of this objective is to gain an understanding of a responsible use of natural resources (water, energy, land). The understanding relates to **developing responsibility** in relation to natural resources use and land use.

Others also mentioned the connections between water and energy consumption saying that a responsible use of water and energy push us towards SD. These lecturers also emphasized a more holistic range of impacts.

Some lecturers believe that this understanding is formed within activities of everyday life. These lecturers emphasised the land impacts caused by energy projects and the importance of **discussing with the students about the responsible use of natural resources (land, water, energy)**. Land is a crucial resource in some Arctic specific areas and the study of the interactions contribute to enhance knowledge about the impacts of extractive activities on land that sustains people's livelihoods. Grete expresses this concern when saying:

GRETE (INDIGENOUS): This is a very important issue; (...) our concern has always been the consequences of energy and minerals for the land.

Finally, some lecturers also expressed the importance of providing the students with understanding of the interactions among natural resources, energy, and environment to equip them to deal with the traditional energy extractive industry and the impacts caused by fossil fuel production on the Arctic and sub-Arctic environment. Renewable energy sources like wind farms also present a range of setbacks in terms of biodiversity impacts which are covered by this course objective as well.

4.2.1.4 Code 4 – to make students aware of natural resources management

In the previous Section (4.2.1.3) the focus was on impacts generated by the interaction of natural resources, energy and environment. This section focuses on the specific skill of managing natural resources as a course objective. Again, the objective is common, but the justifications vary according to the specific context of the Arctic country under study as the resources being managed are different and for different purposes.

Grete (indigenous), a lecturer from the Glacial University pointed out:

GRETE: The focus is wider than energy only. We are very strong in regional and national models for environmental understanding and management. It is not only about energy, but water, land, natural resources and people.

Managing natural resources consists of developing specific skills of assessing the natural assets and managing them in order to use resources in an intelligent and prudent way and make them last to meet the needs of the present and future generations.

At this point, it is not only important to enhance knowledge about the impacts of extractive activities on the environment/land that sustains people's livelihoods but to discuss and provide learning on how to apply their knowledge to manage natural resources in the real world to solve real-life problems. Fossil fuels, renewable resources and land require **management as a skill to be understood, acquired and applied by students.** The courses under study have shown that at Glacial University and at Ice University resources management skills are essential to everybody and they are cultivated in a more practical way through field work and projects in which students need to create solutions by using their skills. There is also a degree of applicability in relation to the course at Borealis University but a more theoretical approach in the course of Juniper University.

During the interviews, there were lecturers that emphasized the importance of equipping students to manage resources in order to deal with economic externalities, i.e. pollution, carbon emissions, and to reduce Greenhouse Gases (GHG) emissions and, consequently, the effects of climate change. Other lecturers were more concerned to equip students how to manage resources in a sustainable way, how to deal with Sustainable Energy, to have basic knowledge and skills to manage energy efficiency technologies and renewable energy technologies.

There were also lecturers that emphasized the aspect of equipping students as professionals to produce environmental assessments on the interactions/impacts of energy, natural resources on the local environment and to be able to work for the energy industry as environmental managers. There were lecturers that reported the importance of this skill for students to work for the oil and gas industry and, other lecturers mentioned the renewable energy industry but emphasized that our present energy system is based and locked in forms of consumption and carbon production that are very harmful. They believe that to deviate from this production and use pattern is very difficult, so we have both consumption and production as systems under review by the next generation of environmental/energy managers. Most of the lecturers expressed their opinions that this kind of skill (resources management) should be more emphasized and truly institutionalized (it is not now, as it is theoretical) as it needs to be taught more systematically by formal education. They also expressed their concerns about the predominance of the technology sector when dealing with resources management by saying that social sciences should be more active in this field to invest more on research, teaching and learning regarding the connections between technology and social sciences in terms of environmental/energy management. It means that the course is not doing it, it is theoretical not practical.

Developing the social sciences side of the matter would contribute to equip students with knowledge and skills to assess and evaluate contexts valuable to be shared with their own communities. It has been done in the courses at Glacial University and Ice University but not at Borealis University and Juniper University.

4.2.1.5 Code 5 – *intergenerational impacts*

In the Glacial University the meaning of SD implies the continuation of indigenous culture and livelihood. It means to have available land for the future or, in other words, available resources for the future communities to be able to survive like the previous generations did. The meaning implies the intergenerational component as follows:

EVA (INDIGENOUS): Sustainable Development for me, means at a very local level the continuation of our indigenous culture and livelihoods in an economically sustainable way but also in terms of available land for the future.

KRISTINE (INDIGENOUS): In our community, I think that Sustainable Development means that we will leave and consume only what we need to, and not excessively and leave the place to our children like we got it from our fathers.

The ability of future generations to survive or, in other words, the intergenerational survival and development of communities is the main objective of their learning experience in the indigenous learning environment. The intergenerational aspect and concern are an integral part of the indigenous culture.

This element is also seen in the interviews of lecturers at Borealis University, Juniper University and Ice University as a component of the formal concept of SD manifested as a concern to balance the use of resources at present and in the future.

EMMA (NON-INDIGENOUS): I think this is a standard notion and it is to ensure that citizens of a particular area understands that development, particularly, the use of resources need to be balanced between the needs of the current generation and the future generations.

RAYAN (NON-INDIGENOUS): the development in the way that protects the ability of future generations to survive and thrive and to develop further without bearing significant consequences of our actions today. Education for Sustainable Development would enshrine individuals that are not aware of the risks of the activities we have today, nor the challenges that may lay ahead (…).

AUDREY (NON-INDIGENOUS): I teach under a broad perspective and it is a term meaning how humans can have a low impact on Earth and living within the limits of ecosystems not using up resources such as carbon-based fuels and looking to Renewable Energy and avoid farming practices that deplete the soil and cause trouble for future population.

'Intergenerational socio-environmental impacts and economic effects' seems to be part of the courses objectives in all the universities under study emphasizing once more the time variable at stake and the attempt to enhance understanding on the natural sciences timeframe.

4.2.1.6 Code 6 – critical thinking

Critical thinking was a common element in all the courses studied. Lecturers from the four universities referred to this element as an objective of their courses directly or indirectly. In aboriginal education critical thinking directly appears as a relevant capability to allow students to understand different cultural perspectives as they are supposed to read literature (social, cultural and historical components) to research and to reflect analytically on themes according to a more evaluative style.

GRETE (INDIGENOUS): the concept of the course is holistic and contextual, which considers all connected aspects of the topic to arrive at a conclusion, including an examination of the own person within the context. Our objective is to provide the students a clear view of cultural, social and historical realities. To build up confidence on the students about knowledge that is not static or under mechanized patterns. Another capability is the capacity of doing research derived from the past and present indigenous knowledge. Critical thinking is an intrinsic characteristic of the course and academism here.

Eva (indigenous) and Arne (non-indigenous) also emphasized the concern on the thinking process and the implications of this continuous exercise as they reported as follows:

EVA (INDIGENOUS): We sit and have discussions; they engage in discussions and ask why? What do you think? How do think things are? Why is the situation like this? And, why that happens? It is participative, democratic, and inclusive; we want to know what our students think about the themes and issues – all opinions are considered, and this is a way of stimulating critical thinking in the students – there is a concern on the thinking.

ARNE (NON-INDIGENOUS): I said the main capability is critical thinking, the students need to understand how to make arguments and what tools and literature they make arguments for and we are not educating consultants we are educating critical thinkers, so we need people who is able to look at academic literature and able to make a reasonable research-based argument.

Nils (indigenous) made a clear criticism of the current curriculum when arguing that the study of ecology is not enough to promote the necessary critical thinking as follows:

NILS (INDIGENOUS): they push the main mission towards ecologists, and they are thinking about something not so useful for the nowadays problems of the Earth – ecology is not enough to promote critical thinking about practical issues of the Earth. The structure is fragmented and, consequently, the content is fragmented because it is separate and contained in specific or under specific titles.

Noah (non-indigenous) also emphasized critical thinking, although indirectly, when pointing out the capability of analysing the options and making decisions to achieve the best solution for a problem in practical terms. Critical thinking seems to be essential when a student has the option of using energy efficiency tools (potentially cheaper) instead of renewable energy solutions (potentially costlier) to provide a real economic and durable solution to a real problem.

NOAH (NON-INDIGENOUS): (...) the main capability is to be able to solve real life problems by using the energy science and knowledge of energy efficiency and renewable energy technology so by creating options and analysing the options to decide which is the best option to solve a real problem.

At Juniper University, in her report, Emma (non-indigenous) mentioned the importance of critical thinking to form educated and informed citizens, as follows:

EMMA (NON-INDIGENOUS): We have one of the highest percentages of students at Higher Education in the world and it is because society, as a whole, seizes a value of having informed-educated knowledgeable citizens. (...) our goal is to encourage people, especially the students, to look at the world in an non-reductionist manner and to learn to defend the opinions they have because it is easy to state an opinion as people can twitter anything, we all have seen that, anybody can twit anything but it does not mean that it is an informed opinion, so if you have an opinion make sure it is informed, make sure you know what you are saying and have a valid reason and fosters respect between the different positions that people end up taking.

4.3 Dissimilarities – features particular to case studies

4.3.1 Glacial University (GU)

4.3.1.1 To develop understanding about values, culture, adaptability, climate change

In general, the acquisition and transfer of knowledge in indigenous culture occurs through elderly members who teach the following generation. They are considered the knowledge keepers and this Traditional Knowledge permeated by values and *praxis* is transmitted by art, language, history, music, medicine, etc, by living in community. For an indigenous student this is the primary source of Traditional Knowledge and the innate mechanism of learning and developing understanding about values, culture, adaptability, climate change, and natural resources.

Indigenous educators tend to relativize the western educational system by promoting an indigenous-centred education and research according to a distinct interdisciplinary methodology, epistemology and pedagogies. It consists in debating and analysing knowledge under the cultural context, truths, beliefs, assumptions and native concepts. It centres on traditional teaching that occurs considering the entire context of what is being studied. Some lecturers at the Glacial University referred to a holistic perspective about the context as a pedagogy in use, where the student is led to consider all the facets and interconnections of the topic under study in order to achieve a conclusion that also considers himself/herself as part of the whole context.

The course looks at themes/issues from an indigenous perspective within an indigenous context. Education at Glacial University follows an indigenous methodology and is developed at several levels. These levels involve a holistic relation with people, with the environment, relations with land as well as relations with ideas. Education takes place in this very particular way, by considering cultural values, challenges, and with focus on how to adapt to change and uncertainty. The course objectives focus on learning about these relations with people, land, environment and ideas. Additionally, the impacts of all these relations are another factor of concern that stimulate field work in order to apply knowledge acquired at home and at university.

Traditional Knowledge is used as a source of information for the Sustainable Development and well-being of local communities. Knowledge has a clear purpose to be produced and applied.

4.3.1.2 To widen definitions of knowledge

To widen definitions of knowledge is perhaps the most particular aspect acquired from the primary data in this research. No other educational institution mentioned this objective. It is particularly important as knowledge at Glacial University is an open chapter. The indigenous perspective captured this dynamic and opened instance of knowledge because knowledge is not static or mechanic. It is dynamic, changeable and it is always possible to enlarge/wide the boundaries of knowledge via Traditional Knowledge.

In terms of Sustainable Development, it seems to be a difficult term to understand and define as the indigenous education conceives of it in a different way. Their perspective about what is sustainable shows a particular understanding of it in the sense that it is not like in the westernized education, but under an Indigenous environmental knowledge about the changes in course in their environment. The indigenous approach and methodology to education and sustainability aims at making people aware of **an identified gap in the understanding** of what is considered sustainable. Education and learning follow the indigenous methodology and take place based on several levels of relationships, with people, with the environment, with land and with ideas meaning that the learning is practical and occurs as a blended experience between the academic and the world reality. TK is researched, compiled and applied to be used as a source of information for the Sustainable Development and well-being of communities.

The acquisition of new knowledge is accomplished within the indigenous context and it comes from the primary data acquired from indigenous people, from the cultural engagement and participatory approach with the community via interviews, apprenticeships, storytelling, cultural observation, filming and photographing where students are researchers creating knowledge according to an indigenous epistemology.

4.3.1.3 To teach the students how to document Traditional Knowledge (TK)

The importance of Traditional Knowledge (TK) to Sustainable Development seems to be significant. At Glacial University (GU) TK is a source of information for Sustainable Development and resource management.

The documentation of TK is not only a way to protect cultural heritage but a form of transmission of knowledge for future generations. It means that the systematic documentation and use of TK contributes to intergenerational knowledge transmission.

To expand and disseminate communities' and societies' knowledge about biological diversity according to their local experiences is a central aspect of the teaching and learning at the Glacial University. Knowing is connected with a theoretical capacity while a skill is linked to a practical

The nature of the Arctic energy curriculum   125

ability. Learning how to document TK involves a process of documentation in which skills turn into knowledge according to Grete.

GRETE (INDIGENOUS): in order to allow the documented knowledge to be part of the living culture, at present and in the future, the knowledge needs to be united with practice (skills), this way the knowledge becomes practice.

Knowledge is not theory it is practice; these two dimensions co-exist at the same time; they are not divided into parts to be exercised in different moments but at once. When students return to their local communities to document TK, they use several methods like audios, drawings, interviews, filming, photographing, as the means to capture specific practices of resource indigenous management systems and the basis for natural capital preservation.

4.3.1.4 To take the idea of conservation of biological diversity and interpret and translate it to the Arctic indigenous context

Sustainable Development is embedded in the course despite not being called 'sustainable development' but biological diversity. When students go and document biological diversity through the practices adopted in his/her area and later share this experience in class, this widens definitions of knowledge. Along the circumpolar Arctic there are a myriad of communities with their ancient practices which are shared, interpreted and translated into the Arctic indigenous context. The shared experiences in class based on documented TK provide a common basis for understanding the region and the surrounding world.

Sustainable Development is translated by conservation of biological diversity, in a way that the conservation of life in the physical environment (geography) is an intrinsic premise for the conservation of people's culture and livelihoods in relation to human environment (human = social geography). These are two in-dissociable elements, meaning that when a student researches the physical environment, he is also studying the human environment. It is just one thing because practical knowledge comes from people.

4.3.1.5 To provide knowledge that can be disseminated in communities

Documented TK has the purpose of being shared and accessible to communities. This is the end of the process and the purpose of equipping students with knowledge that can be shared with their own communities and with the multicultural group. The purpose of this learning and field work, according to Eva (indigenous) is to be used and be helpful for sustaining

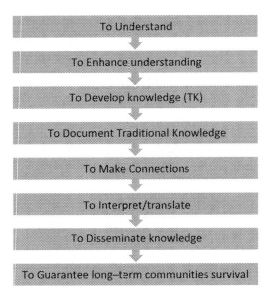
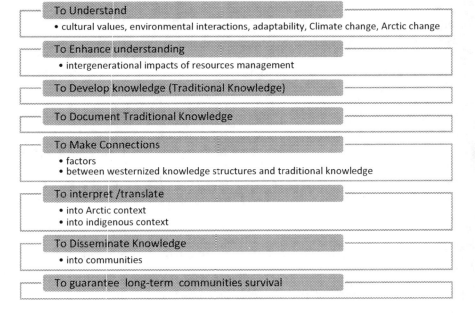

Figure 4.1 General (top) and Detailed (bottom) Indigenous Education Pattern – Course Objectives of Glacial University (GU).

indigenous societies. This learning experience promotes Sustainable Development as per indigenous perspective:

EVA (INDIGENOUS): The lectures basically give the students a combination of options of different topics. We have a lecture, for example, focusing on very technical aspects of Biological Diversity but also students have lectures about Traditional Knowledge and how to document Traditional Knowledge. What you can use Traditional Knowledge for. How Traditional Knowledge can be used and be helpful for sustaining indigenous societies.

To document TK on biological diversity is an important process with a specific purpose of sustaining indigenous communities but it also highlights the community role of efficiently managing the knowledge (see Figure 4.1).

4.3.2 Borealis University (BU)

4.3.2.1 To learn the connections among factors

At Borealis University (BU) the emphasis is to understand Sustainable Development by identifying connections between many factors. This is a multifaceted and interdisciplinary concept and the main point of geography is to show the students these connections

OVE (NON-INDIGENOUS): (…) the students come and are interested in a broader view because of different kinds of jobs (they want to be prepared for jobs). This is Sustainable Development; you need to see the connections between many factors and the main point of geography for us is to see these connections.

The connections occur at different levels. Connections among different modules, contents mainly relating population, globalization, environmental impacts, energy and climate change. All these components together contribute to the understanding of SD. In this particular course, energy is studied through the theme of climate change.

4.3.2.2 To show the importance of local participation

Participation is another fundamental aspect of the course at Borealis University (BU). It is translated into public participation or societal impact.

Randi (non-indigenous) argues that the course is not only for educating people for academic positions or for jobs, but it provides opportunities for students to have a public participation as community workers and activists.

Local and global impact of knowledge is an important objective pursued by the programme.

4.3.2.3 To apply knowledge in international geographical areas

The student can apply their knowledge on SD and energy when returning to their countries – knowledge application into practice locally according to geographical characteristics – geographical customized energy and development solutions

4.3.2.4 Employability vs. knowledge application – dilemma (to train professionals but not only to apply knowledge in career)

The dilemma of employability versus knowledge application is a common feature in the westernized education. This trace was realized mainly at BU. The main concern for students is to have a job after finishing the course, but not the application of their learning.

Lecturers also emphasized this point as a course objective. Some of them reported that the courses aimed at equipping students with skills for employability/jobs, for accessing international job opportunities in different sectors, i.e. industry, NGOs, activism, academy. At BU there were reports from lecturers demonstrating a concern about knowledge application in life, but it seems to be a theoretical discussion without practical evidence, and effects of this dilemma can be seen in the curriculum when orienting geographical discussions to the field of ecology. One specific objective in the targeted course at BU is to equip students for future employment and accessing funding (see Figure 4.2).

4.3.3 Juniper University (JU)

4.3.3.1 To have informed and educated knowledgeable citizens – global citizenship at JU

In Juniper University (JU), some lecturers reported that they had one of the highest percentages of students at HE in the world and because their society as a whole seized a value of having informed-educated knowledgeable citizens.

The word 'citizen' or 'global citizen' was used in some occasions but it was not possible to verify with certainty if the concept of global citizen was really understood or if it was only an expression being used. It was clear that the national educational policies are oriented to stimulate formal education, but it was not clear if a true 'education for global citizenship' as per the concept of Delors et al. (1996), Hill (2000) was being practiced. Taking into consideration the apparent cultural tension between indigenous and non-indigenous and reports from students about the lack of meaningful use of the concept of SD, which will be discussed in more detail in Chapter 8, this point brings doubt if this notion is fully understood and applied.

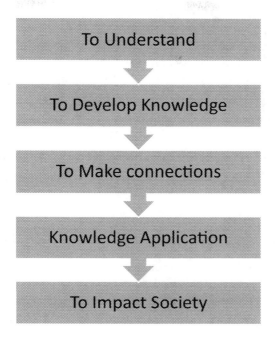

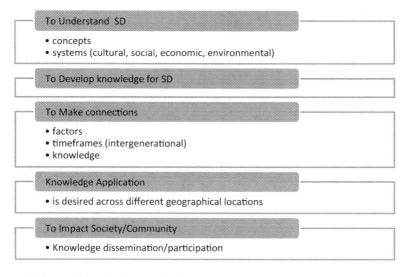

Figure 4.2 General (top) and Detailed (bottom) Pattern – Course Objectives of Borealis University (BU).

4.3.3.2 To sustain and to enable SDL

It is not just a matter of learning the concepts but how to maintain SD and energy literacy in continuous operation and evolvement. This particular objective seemed contradictory as SD is a separated field of study and energy literacy was not a well-known notion at JU as per the interview reports.

4.3.3.3 To teach students about energy policy, risk management, CSR, international business

The combination of geography and business strategy in relation to natural resources management is a strong characteristic of the course. It emphasizes the point of preparing students for being managers in different industries.

4.3.3.4 To enhance understanding on energy dimensions (business, government, economics, policy, geography, finance risk, etc)

Lecturers reported that students undertaking the course are able to look at a particular issue from a multifaceted perspective. The economic, political and financial aspects are highly emphasized in the energy study, confirming the trend to introduce the student to the business and management environment of different industries.

4.3.3.5 To prepare students for the energy industry

To train future energy industry professionals was another particular objective reported at JU. This corroborates the previous objectives mentioned on 4.3.3.3 and 4.3.3.4 meaning forming students to operate in the energy industry or training energy workforce (see Figure 4.3).

4.3.4 Ice University (IU)

4.3.4.1 To provide energy knowledge and information for a lifetime

At Ice University (IU) some particular objectives were observed from the lecturers' reports like 'to provide energy knowledge and information for a lifetime' or (life-long learning); to provide education for promoting positive changes in energy practices (practical applied knowledge); to solve real life problems; to create solutions to energy problems; to provide real life learning experience (practical projects); to equip students to deal with sustainable energy; to provide knowledge on energy efficiency; to provide knowledge on renewable energy; to provide people with the means to make educated and well-informed decisions in their daily lives.

The approach at IU is more practical and the focus is knowledge application to solve problems of the everyday life (see Figure 4.4).

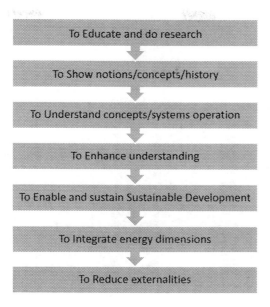

To educate and do research
- western concepts and systems

To show notions/concepts/history
- Concept of SD and history of the concepts evolution

To understand concepts/systems operation
- economic and cultural systems
- production and consumption systems
- energy systems/fossil fuels system

To enhance understanding
- concepts and systems

To enable and sustain Sustainable Development
- concepts

To integrate energy dimensions
- business, economics, government, policy, geography, finance, risk, management, industry

To reduce externalities
- reduce impacts (i.e GHG emissions, environmental risk assessment

Figure 4.3 General (top) and Detailed (bottom) Pattern – Course Objectives of Juniper University (JU).

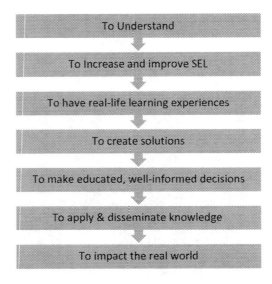
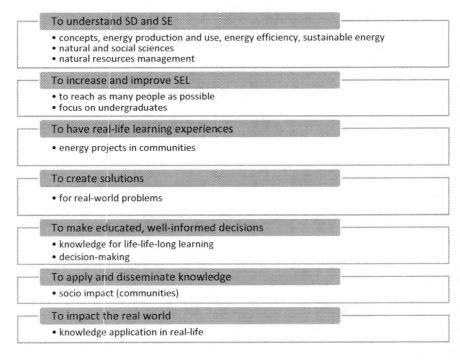

Figure 4.4 General (top) and Detailed (bottom) Pattern – Course Objectives of Ice University (IU).

The nature of the Arctic energy curriculum 133

4.4 Key objectives of the courses as per lecturers' interviews

The nature and characteristics of the curriculum are highly determined by the objectives it aims to accomplish. Curriculum is usually viewed as an instrument of change and this section outlines the characteristics that define the course curriculum in each of the studied locations as per the coding process. Figure 4.5 illustrates the heterogeneous feature of the objectives in the case study universities curricula.

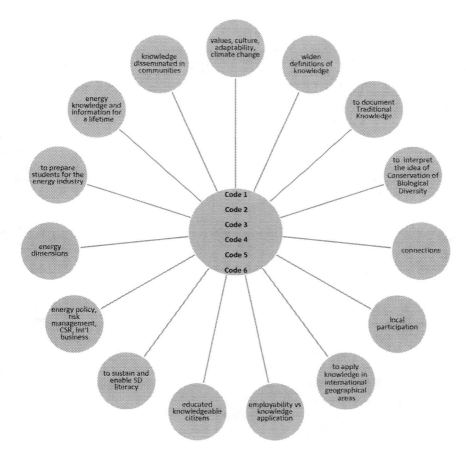

Figure 4.5 Key objectives of the courses as per lecturers' interviews and coding process. The codes that were agreed across all participants (1–6 from Table 4.2) appear in the centre, but this figure is to show the diversity of other objectives mentioned by the faculty members (lectures).

Based on the common and distinct characteristics of each studied location under the criteria regarding course objectives, it is possible to categorise the objectives in the case study universities as follows:

134 *The nature of the Arctic energy curriculum*

a. Glacial: Developmental, Practical Applied Knowledge
b. Borealis: Theoretical and Developmental Knowledge
c. Juniper: Theoretical Formal Knowledge
d. Ice: Developmental, Practical Applied Knowledge

Table 4.3 Features of Education in the case studies shown as a continuum from theoretical to applied knowledge

University	Features of Education
Juniper Univ. (JU)	Theoretical Formal Knowledge (no concept, no use of energy literacy) not a known notion
Borealis Univ. (BU)	Theoretical and Developmental Knowledge (energy literacy (not official term) but the notion is a component of ESD)
Glacial Univ. (GU)	Developmental, Practical Applied Knowledge **(energy as an implicit element of Conservation of Biodiversity)**
Ice Univ. (IU)	Developmental, Practical Applied Knowledge **(sustainable energy literacy as an autonomous course as a realization of ESD)**

Surprisingly, the coding process applied to the lecturers' interviews findings revealed more dissimilarities than the common points verified by the codes from 1 to 6 demonstrating a high level of heterogeneity (see Figures 4.5 and 4.6) in the HE system in the Arctic. The curriculum objectives, approaches and content were highly determined by the cultural and economic systems in place in the four case study contexts, demonstrating that modes of learning are different, and they vary from a more theoretical to a more experiential and practical levels.

The findings from academic staff members' interviews presented in this chapter contributed to uncover important components concerning the objective of the courses and the nature of their specific curricula according to academic staff's teaching experience and practices in the case study locations. Based on the data collected from the four institutions it was possible to identify the key objectives of the courses as per lecturer's perspective including similarities that were codified as per codes 1 to 6 and a higher number of dissimilarities showing particular characteristics of the case studies making it possible to categorise the case studies into levels that vary from a more theoretical to a more practical learning mode (see Figure 4.6).

Finally, the findings contained in this chapter inform the discussion about the multidisciplinary and multicultural perspectives of a sustainable energy model being addressed by Higher Education curricula in the Arctic nations as it is shown in Chapter 10 (Sections 10.2 and 10.3).

The nature of the Arctic energy curriculum 135

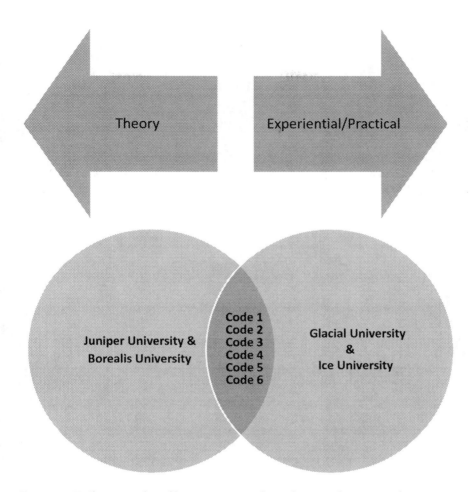

Figure 4.6 Different modes of learning varying from theoretical to practical.

References

Bruner, J 1996, *The culture of education*, Harvard University Press, Cambridge, MA.
Delors, J, et al. 1996, *Learning: the treasure within*, UNESCO, Paris, p. 37. Report to UNESCO of the international commission on education for the twenty-first century.
Hill, JD 2000, *Becoming a cosmopolitan: what it means to be a human being in the new millennium*, Rowman and Littlefield, Lanham, MD.
Lave, J & Wenger, E 1991, *Situated learning: legitimate peripheral participation*, Cambridge University Press, Cambridge.

Mertens, DM 1998, *Research methods in education and psychology: integrating diversity with quantitative and qualitative approaches*, The University of Michigan. Sage Publications, Thousand Oaks, CA.

Parlett, M 1981, 'Illuminative evaluation', in P Reason & J Rowan (eds.), *Human inquiry: a sourcebook of new paradigm research*, John Wiley & Sons, Chichester, pp. 219–226.

Rogoff, B 1990, *Apprenticeship in thinking: cognitive development in social context*, Oxford University Press, New York.

5 The structure of the Arctic energy curriculum in the case studies

5.1 The structure of the energy curriculum

5.1.1 The structure of the energy curriculum – the most influencing factors and distinctions

This chapter outlines findings derived from the documentary phase (phase 1) and academic staff interviews (phase 2) of the research by presenting a consolidation of the elements contained in the syllabuses (module document containing learning outcomes and additional module information) of the targeted courses in the case study institutions. Energy studies and human geography/sustainability programmes were identified as an opportunity to critically examine if the process of energy transition is approached in the curriculum. Along this path, it was identified that Higher Education (HE) has a much more relevant role than only acknowledging this process of transition. HE curriculum presents the potential to insert students into a sustainable energy vision in practical terms. The relationship between energy and human geography is intrinsic and highly complex in the Arctic; as van der Horst et al. (2016) observes, cheap and abundant energy determines the type of geographical constraints that influences societies to deal with space (land), climate, sustainability, and other societies (globalization). This is a central concern of this section: to understand the structure of the energy curriculum in the Arctic HE according to the most influencing factors and dissimilarities identified in the curricula of the case study institutions. This is the reason why the curriculum is the emphasis of this chapter because the focus was to understand the structure of the energy curriculum in the Arctic considering the human factor and how energy influences societies. It was also important to understand the structure, content, approaches, and pedagogies based on the documents and staff interviews.

Curriculum structure depends on curriculum design that according to Fish and Coles (2005, p. 17) can follow a Product model, a Process model, or a Research model. The first is lecturer-centred and depends on the lecturer's knowledge transmitted to the students that receive the knowledge passively; the second is student-centred and a favourable environment is provided by the lecturer to the student to operate his/her learning; and the third model is student-centred, and lecturer and student participate in the learning in a more collaborative way. The second and third models are aligned to constructivism.

Curriculum structure also relies on the curriculum frameworks to develop an understanding of the knowledge, skills, generic, and discipline-specific capabilities required by the students. The curriculum structure of the courses in the four case studies relating components of energy and human geography/sustainability were very diverse depending on the geographical and cultural context and do not present similarities to specific energy curriculum developed for undergraduates studying for roles in the energy industry.

In terms of curriculum structure, there were three most influencing factors identified in the curricula of the four institutions: the curricula involve research at different levels (Healey 2005); the curricula are interdisciplinary; and 'resources management' was as a common content component identified in the four curricula structures.

5.1.1.1 Research as a structural course component

The courses in the four universities targeted by the study have research as a common feature. In general, the curricula in the four case studies

Table 5.1 Curriculum characteristics based on models of the research-teaching nexus by Healey (2005)

University	Curriculum characteristics
Juniper Univ. (JU)	Curriculum places emphasis as much on understanding the processes by which knowledge is produced in the field as on learning the codified knowledge that has been achieved. Developing students' research skills (explicitly, in addition to other disciplinary and generic skills).
Borealis Univ. (BU)	Curriculum is structured around subject areas and the content selected is directly based on the specialist research interests of teaching staff. The opportunity to work on research projects alongside staff.
Glacial Univ. (GU)	Curriculum is designed based on systematic inquiry research. Using assignments that involve elements of research processes.
Ice Univ. (IU)	Curriculum is largely designed around inquiry-based activities, rather than on the acquisition of subject content; the experiences of staff in processes of inquiry are highly integrated into the student learning activities. Using teaching and learning processes that simulate research processes.

were structured in each institution as a collaborative effort among lecturers that are also researchers in their expertise areas. In each studied institution, the course stimulates the development of research skills in the students but in different ways and at different levels. The lecturers reported sharing their experiences to compose the content, methods of delivery and assessments by connecting learning and research (see Table 5.1) through themes associated to natural resources, climate change, cultural well-being, society, environment, biodiversity, land use, and sustainable energy.

5.1.1.2 Interdisciplinary structured courses

Another influencing factor is the interdisciplinary characteristic of the courses in each studied location. Depending on the local context, certain themes are more or less emphasized in terms of research, and a combination of options based on interconnected topics like climate change, land use, energy production and use, environmental impacts, biodiversity, and societal issues is offered. An interdisciplinary approach among subjects also happens considering that one course can combine modules like anthropology, geography (physical and human), sciences, history, art, society, and culture. It means that a student can study energy combining different intersections, like science, social environment, and the physical environment.

5.1.1.3 Resources/environmental management as a common theme across the case studies main course structure

The third common pattern in terms of structure among the courses studied in the four different locations was the management factor. The courses present an intrinsic thematic concern on how to manage local natural resources and the local environment regardless of the type of resources in question (land, energy (oil and gas, renewable energy), sea (fisheries)). In one location the important resource can be land or the sea; in the other it can be wind and solar power; in another it can be oil or the water pollution. The management of impacts is a common concern and content conveyed in the courses analyzed in the studied institutions.

IVAR (INDIGENOUS): In our teaching course we have this component of Sustainable Development (SD) related to energy, natural resources and how to manage them.

There were opinions in relation to management in the sense that there are still several aspects not approached by the Higher Education curriculum like over-consumption and carbon emissions. Some discussions exist, according to lecturers, but they do not seem to be deep enough to address the issues in practical terms. Despite being a common theme, it is approached according to different levels of depth and *praxis*.

5.1.1.4 Distinctions in terms of structure, content, and approaches

5.1.1.4A GLACIAL UNIVERSITY (GU)

According to the findings from the interviews with lecturers from the Glacial University (GU), the course was structured in a traditional way (academic emphasis as per Western style), but it has very untraditional components according to Eva and Grete (lecturers). In general, the course follows academic standards (inspired in Western quality credentials); it is linear and holistic in a combination of place-based learning and distance learning, delivery elements that determine the type of learning activities, assessments, and the course dynamics. The students come from all over the circumpolar north and it is a multilingual and multicultural course. The untraditional components refer to the specific course objectives reported in Chapter 4 among them, the documentation of Traditional Knowledge (TK) through an indigenous-centred research methodology based on responsibility, intelligences, and primary data collected directly from indigenous people. An example of untraditional components are the cultural activities the group experience together at the education institution and the 'real-life learning experience assessments' based on cultural immersion, apprenticeships whose outcomes are disseminated in the communities across the Arctic, making the students multipliers of the co-created knowledge.

On one hand, the academic scientific knowledge is a contact point to a traditional style; on the other hand, the course is structured based on the experiences of lecturers and students as co-managers of knowledge within a cultural context. The idea was to develop a course as a collaboration among researchers of the institution and with a partner research centre that operates within communities.

The structure and content were designed based on researchers' shared experiences. Basically, the course structure, content, approaches, and pedagogies grew out of their experiences in documentation of TK. The participatory approach within the cultural context gives the students the freedom to connect to their areas of interest; at the same time, it bridges curriculum content between departments.

The students come to the campus for a period of time, where group activities take place in a very intensive daily regime of lectures from 8 am to 5 pm that provide the students a combination of options of different topics related to the main focus of technical aspects of Biological Diversity (equivalent to Sustainable Development (SD) in Western education). The interdisciplinary topics are studied under the viewpoint of

TK, on how to document TK, what to use TK for, and mainly how TK can be used for sustaining indigenous societies. The course is structured based on an indigenous perspective of natural resources, environment, and society.

The students at the Glacial University (GU) have practical assessments consisting in returning to the hometown communities in order to assess and document TK in relation to the selected aspects of biological diversity (based on the Convention of Biological Diversity – CBD) After concluding their work, they return to the campus and present it to the group (staff and students, multicultural group). This is about socialization, connecting to each other, sharing experiences to build up future collaborations, and disseminating knowledge to their communities.

The students work on a project to document TK in their communities and they are incentivized to do interviews; videos (filming), multimedia, or other method they believe is more appropriate to document (record) the knowledge (TK) from the field. They basically choose a method and a narrow topic to document biological diversity by TK with the assistance of a mentor throughout the period.

The students also have discussions in small groups in a seminar style conducted by tutors stimulating discussions supported by the use of Power Point slides and other multimedia resources. In general, there is broad use of figures, pictures, and texts to diversify and make the human encounter more dynamic, as it is not the staff style to stand in front of the students talking for 2 hours. The curriculum, pedagogies, and approach are more interactive and holistic by using audio-visual method more appropriate to an indigenous curriculum. Staff reported that tutor and student usually sit together and have discussions; they engage in discussions and the tutor asks questions like: Why? What do you think? How do you think things are? Why is the situation like this? And, why that happens? This is a very reflective approach.

It is not only very reflective, but it is participative, democratic, and inclusive, as the staff constantly searches to know what the students think about the themes and issues that are taught and discussed – all opinions are considered and this is a way of stimulating critical thinking in the students – there is a great concern on the thinking and reflection, but it is mainly about values, culture, awareness, environment, change and adaptability, and science and culture. Teaching and learning are a cultural experience, as lecturers and students participate in the learning (applying scientific knowledge acquired in class) through cultural daily practices.

An indigenous curriculum is adopted in the Glacial University (GU). This is the nature of curricula there. The type of curricula informs how TK can be used and be helpful for sustaining indigenous societies.

5.1.1.4B BOREALIS UNIVERSITY (BU)

The most important characteristic of the curriculum at the Borealis University (BU) is to make students understand the importance of participation to the study of geography and SD. The course is modular and research skills are developed through literature reviews and funded projects. A more participative approach to stimulate critical thinking is an important feature of the course structure motivating participative pedagogies.

ARNE (NON-INDIGENOUS): The course is structured to make the students to understand the importance of participation to this process – participation to Sustainable Development.

The staff at Borealis University (BU) are also of the opinion that SD relates more to ecology, but it should relate more to geography studies. According to Nils (non-indigenous), 'Sustainable Development (SD) and ESD is something in ecology but it should be something in geography':

NILS (NON-INDIGENOUS): (...) ecology is not enough to promote critical thinking about practical issues of Earth. The structure is fragmented and consequently the content is fragmented because it is separate and contained in specific or under specific titles. Geography course is fragmented in different segments instead it should have a more comprehensive and connected approach.

He also adds that geographers should have a more holistic viewpoint to exert a more useful role in resolving the current problems of the Earth, but SD has been the mission of ecologists and he argues that the curriculum tends to emphasize ecology because of the better job opportunities it offers, but ecology, in his view, is not enough to address the Earth complexities. He also was emphatic when saying that, consequently, the course structure is fragmented and compartmented not serving the purpose of geography:

NILS (NON-INDIGENOUS): Geography is much more about ecology and it is reflected in job opportunities because as an ecologist there are more chances to find a job in geography. Basically, environment is about ecologies and geography is about maps and cities – it is not enough to approach and address the complexity of the current earth issues. As a geographer it is more difficult to find a job because there is not the right mentality about the mission of the geographer. The objective and structure are fragmented and consequently the content is fragmented and compartmented and contained in specific and separate and specific titles.

Other interviewed lecturers observed that the combination of the modules in human and physical geography, dealing with sustainable energy, contributes to the learning of SD. In this regards the curriculum is modular and the content is interconnected. Different modules present interconnected content, for example when discussing about population, it is connected to globalization and climate change. Energy is very much linked to climate change and it is studied under the module of climate change.

SD is one of the key concepts of the course and it is studied in a reflective style through the application of participatory exercises in the class. Lecturers emphasized their intent not to reproduce knowledge for good grades but to promote reflection. Application of concepts into practice is something much desired at Borealis University (BU).

RANDI (NON-INDIGENOUS): I want them to work differently, I want them to try to find what they want to do and what is the consensus and what the other students are contesting, asking, and how they reflect, so I think the way I teach is more participative.

Application is desired by the institution when promoting field research and the students' participation in funded international energy projects, but no specific figures were provided in order to measure the level of current participation in projects. The interviews suggested that participation is desired but needs to be enhanced.

5.1.1.4C JUNIPER UNIVERSITY (JU)

According to the head of department of the sustainability programme, energy is an implicit part of their lives because their whole economy is based on energy, for this reason she argues that:

EMMA (NON-INDIGENOUS): It is not explicitly but we live in a place where energy is an implicit part of our lives. Our whole economy is based on energy (...) so there is no explicit discussion of energy in our programme, but it is used as examples throughout every class.

According to Emma, the course is structured in a way that it focuses on two particular areas or on the intersection of health and environment. These two areas represent the social environment and the physical environment. The curriculum is modular and with a specific focus on the intersections 'where people live and their health' and on 'environmental management'.

The course is largely lecture-based and covers topics related to economics, agricultural revolution, environmental issues, health, and quality of life, inclusive energy. Climate change is a sensitive topic, as there is some scepticism particularly amongst those individuals from the energy industry. According to Rayan,

RAYAN (NON-INDIGENOUS): I am very cautious at the beginning to not have a very strong environmental message and this is critical – sustainability and climate change are sensitive subjects for some audiences.

The approach has a systemic element – systemic thinking is a component of the course structure or the course approach and it is reflected in a discourse about the economic system and the socio-environmental impacts of the current economic system in which we are inserted:

RAYAN (NON-INDIGENOUS): (...) environmental impact is part of our economic system and this is essentially a by-product that we haven't managed to address including when we go to more complex issues as power.

The programme is modular, and students can combine interdisciplinary modules according to their interest. As the students are mixed, meaning that we have indigenous and non-indigenous, there are differences in the modules these two different groups choose. Indigenous students tend to opt for modules related to native geographical history and studies versing on historical dispossession and looking at specific extractive projects when industrial capital approaches upon people's territories as well as the relationship between these actors.

On the other hand, non-indigenous students tend to choose more energy industry-related modules including energy finance, energy law and policies, energy markets, environmental policy, energy business, governance, and relations that can be other contents of the course depending on the combination of modules the student elects.

Finally, Zach (non-indigenous) reported the importance of awareness in relation to SD and energy. In his interview, he said that he taught energy and it is part of SD, but he also emphasized that sustainability programme is separate from energy.

ZACH (NON-INDIGENOUS): I guess we have a sustainability program separate from energy. Probably it is a good question for them. I teach energy and it is part of Sustainable Development, but I don't think what I am doing is Education for Sustainable Development. I think energy literacy is important to understand the policies, the implications, the markets, and about what may contribute to be a path towards Sustainable Development.

The structure of the course keeps two territories: one is energy and the other is sustainability. The energy territory covers business, markets, policies, and the industry; the sustainability territory discussed environmental management, environmental impacts. There seem to be a line dividing these two subjects and they do not mix. On one hand, there is recognition of the importance of ESD and energy literacy; on the other hand, to cross the boundaries and mix energy and sustainability is the cause of an implicit tension. The focuses of these two areas are different. One side focuses on the energy industry, production, and demand, but the other focuses on socio-environmental impacts and management reflecting historical, cultural, and economic tensions.

5.1.1.4D ICE UNIVERSITY (IU)

The structure of the course in Ice University (IU) comprises core subjects and electives that allow students to specialize in the areas of their interest, but the course structure is modular and focused on sustainable energy and energy literacy. These are the central elements of the course. The central component, object, and approach of the course are to promote basic energy literacy to the highest number of people possible. The second most important element is to promote real changes with the application of energy knowledge by reaching the largest number of students as possible, as Noah observes:

NOAH (NON-INDIGENOUS): my approach is to really promote the basic energy literacy because with literacy we can make the biggest changes.

NOAH (NON-INDIGENOUS): that is my approach, because if I focus on students with higher degrees, masters and PhDs, I would be reaching a small population, by focusing on undergraduates I can work with so many students as I can.

Students that opt to this course want to be more knowledgeable about energy. The focus is on undergraduate students because the programme aims at reaching and disseminating knowledge and information to the largest number of students as possible. The course aims at equipping students to the practice, the day-to-day life, and work. Energy literacy is very relevant because students will be dealing with energy in their occupation no matter what they do for a living. In fact, the staff believes that energy literacy is important to everybody not only for people that work directly with energy as Noah observes:

NOAH (NON-INDIGENOUS): a school principle or a campus director, these are people whose jobs are not directly related to energy, but they are dealing with energy, they are making decisions about energy in relation to their buildings; they hold responsibility to deal with curricula for their schools.

In relation to the content, it is focused on the basic energy literacy, energy efficiency, renewable energy, and energy science with a component of socio-environmental impacts.

The course attracts a variety of audiences. The students are mixed: indigenous and non-indigenous and at different age ranges, as youth and more mature students are part of the group regardless of occupation, background, and career interests. It is particularly attractive to secondary teachers who might be teaching subjects related to environmental topics in their classes. In general, the lectures are video recorded to engage the students to watch in advance, to review the content, to share ideas, and to have additional conversations making sure they clearly understand the content and can share their experiences with the tutor. Students are also asked to provide feedback on how the lectures went or if it did not work for them, as the groups are very heterogeneous and students have different ways and needs of learning. Feedback is used to constantly improve and adapt the course in order to disseminate accurate energy and environmental information that, in the case of the secondary teachers, ends up being transmitted to secondary students.

The lecturers emphasized the benefits of teaching using multimedia, videos, audio recordings, and power point presentations that proved to stimulate student engagement in energy literacy:

OLIVIA (INDIGENOUS): In this case we use recordings and so it was good because the students are not experts in sustainable energy and this helps them to overcome any kind of reluctance in dealing with the information, they are not familiar with, so if they have to teach about it (when the groups are formed by secondary teachers). So, these are the benefits of teaching using the videos and to prepare presentations and when teaching the secondary teachers, they can prepare the presentations and modify and use them in class if they want.

They branch out based on students' interests; in other words, students can combine courses according to their interest on energy subject. The course always starts with general concepts and evolves to more technical paths and practical approach focused on energy efficiency, clean energy, and SD.

There is a clear intent in making the course even more practical as Thomas emphasizes:

THOMAS (NON-INDIGENOUS): I want to do more (...) this means to provide the students hands-on opportunities – completely practical approach (hands-on).

Students search for the course as they intend to immediately apply their knowledge in the contexts of their daily lives. In this sense, the course assessments (mid-term and end-of-term) are related to their real-life needs and students' real-life projects. Students are asked to create and work on projects dealing with real problems they want to have solved in their daily life experience, for example, issues related to off-grid energy generation/use, house insulation, ventilation, energy conservation and efficiency, assessments related to the best renewable energy sources, and technology according to the geographical context, community training, etc.

In general, students design projects (and present results in class) related to their participation in smart-grids/mini-grids of renewable energy for urban and remote Arctic communities; projects aiming at assessing energy consumption levels in farms/houses/offices; home/building insulation materials and methods; energy-efficient devices and appliances; economic and technological options of off-grid energy generation/use in communities and impacts when taking part of community projects, when disseminating their knowledge to train secondary students, or when training community members how to deal with renewable energy.

5.2 Capabilities expected to be developed in students as a result of the objectives, structure, approaches, and pedagogies

5.2.1 *Glacial University (GU)*

As the objective is to provide the students a clear view of cultural, social, and historical realities and to document Traditional Knowledge, the students develop capabilities related to confidence about the importance of their knowledge and the capability of adaptation to a non-static knowledge, a changeable knowledge that does not work under mechanized patterns.

EVA (INDIGENOUS): You should see our students from Monday one week till Friday second week, they might come here very shy, first time they are ever outside their home village and by two weeks they are presenting like professional academics. It is something about building their confidence about their own Traditional Knowledge about the topics that exist in the International sphere, for example biological diversity, about the techniques you can use to communicate this knowledge and, of course, the techniques you can use to document your own Traditional Knowledge to make it applicable and available to broader governance.

Another capability is the capacity of doing interdisciplinary research derived from the past and present indigenous knowledge and how to document this knowledge by using indigenous methodology and inquisitive mind-set. Deep reflection and critical thinking are intrinsic characteristic of the course and academism at Glacial University (GU).

GRETE (INDIGENOUS): the concept of the course is holistic and contextual which considers all connected aspects of the topic to arrive at a conclusion, including an examination of the own person within the context. To build up confidence on the students about knowledge that is not static or under mechanized patterns. Another capability is the capacity of doing research derived from the past and present indigenous knowledge. Critical thinking is an intrinsic characteristic of the course and academism here.

Another important capability identified in the interviews is knowledge application. Specific cultural activities are held to discuss and to show how to apply their knowledge by employing real-world examples and empirical workshops (like the kick-off exercise in the experimental kitchen where the indigenous students share their cultural roots and community practices by preparing a traditional dish from their communities and talking about environmental ancient practices related to local social, environmental, and economic activities). Additionally, the content and format of assessments also make the application an important capability at Glacial University (GU). This capability is also linked to the ability to connect the theory and the real world and establish the connections and the interactions between natural science and social sciences:

IVAR (INDIGENOUS): Capability would be to increase their understanding of how things are depending on the nature, basic understanding of the chemical interactions if you are discussing oil extraction or mining or any kind of material or radioactivity and the impacts they cause.

The students need to receive the basic knowledge in order to be equipped to understand basic concepts and make co-relations with other disciplines and the Traditional Knowledge. They learn chemistry that represents the basics of Western knowledge and they can make the connections with Traditional Knowledge. They can read the book and they can understand but they need the knowledgebase. This perspective means that (the same view as in Ice University course) knowledge needs to be basic, accessible, and intelligible to everybody. It needs to reach everybody to mean change.

Another fundamental capability is the knowledge about identity, values, and awareness about the cultural differences. The students are prepared to understand diversity and cultural differences and they can apply this knowledge in their daily lives. When students go back to their communities, they see the value in their own culture and they also teach that to others.

KRISTINE (INDIGENOUS): What we provide here, what I got from my education here, was more knowledge about identity, values and awareness about these cultural differences. We prepare the students to understand diversity and cultural differences. When they go back to their communities, they see the value in their own culture and they also teach that, and it is a healthy approach to counteract the mainstream media discussions about sustainability. The students are able to teach their communities about Sustainable Development. I don't think that we are so aware about our role or the role of our knowledge, of our values, but when we are taught more about it, we get more awareness about our own values.

KRISTINE (INDIGENOUS): When you are forced to reflect on where you come from, what you value in life, how the development is going, how the community is going, when you are forced to reflect on it you grow awareness and you can change how they think about i.e. Sustainable Development.

The students are able to teach their communities about SD – students are learners and teachers or disseminators of knowledge, information, and awareness. The students understand the knowledge and their role in face of knowledge and values. This understanding, critical reflection, and dissemination make them develop/grow awareness and be agents of change, change-makers in their cultural and natural environment.

5.2.2 Borealis University (BU)

The objectives and course structure at Borealis University (BU) privilege critical thinking and reflectiveness as the main capabilities as staff expressed in some interviews that they are educating critical thinkers. As Arne says:

ARNE (NON-INDIGENOUS): The main capability is critical thinking; the students need to understand how to make arguments and what tools and literature they make arguments for and we are not educating consultants we are educating critical thinkers, so we need people who is able to look at academic literature and able to make a reasonable research-based argument.

Reflectiveness, criticality and values component are also seen in Nils report:

NILS (NON-INDIGENOUS): Some of them, for example, are more on the side of liberalism others more on the side of socialist ideas, and I explain to them that when we say Sustainable Development we need biodiversity; biodiversity is not just about the environment, it is also about biodiversity in human landscape, in my culture landscape, and for having biodiversity you need democracy, you need liberalism and they like that, and for Sustainable Development you have to fight against thirsty, you need to fight against inequality, because you need equality for Sustainable Development, some students like Sustainable Development because of sociology and I explained to them that if you want to change your behaviour and your personal lifestyle, and live on Sustainable Development you have to change your personal indexes from quantity indexes to quality indexes more comfortable home, not bigger home, more comfortable car not more expensive car; you have to change this to have a lifestyle adapted to Sustainable Development but by having quality indexes in lifestyle it shows your mentality about life;

Another relevant capability is to think in an interdisciplinary way. The students are taught and stimulated to learn how to make connections between SD and other disciplines like sociology, psychology, climate change, energy. Nils (non-indigenous) argues that this knowledge is not just about the environment, it is also about biodiversity in human landscape in culture landscape. He adds that 'for having biodiversity you need democracy you need liberalism, you need equality for Sustainable Development. You need a deep review on personal and collective consumption behaviour needs and patterns'.

Finally, Ove (non-indigenous) and Randi (non-indigenous) observed that students really appreciate studying SD under this interdisciplinary perspective and a great number of students opted to work on practical projects related to SD and present their final outcomes to the group (staff and students).

OVE (NON-INDIGENOUS): I think they have shown after all, after all these modules, they have chosen projects on Sustainable Development, On December 15th, they presented their projects on Sustainable Development to the group.

RANDI (NON-INDIGENOUS): The core course I am teaching provides a manual of different ways to think about development and for them to identify what is the most appropriate content for them to use in the book. In the methodology classes in geography there are sessions on GIS and on quantitative methods, i.e. doing surveys and using questionnaires. I am responsible for teaching the other methods, focus groups discussion, interviews, photography, there are so many different things, it is a manual of different techniques where you have to be very much present in the field and have contact with the real facts and realities on the ground and then, knowledge.

RANDI (NON-INDIGENOUS): I think the capability I hope to build in them is to be sensitive to what happens on the ground; to see the poverty in their own societies, to pay respect to the people that you study and also, as said, you have a responsibility to disseminate knowledge in a very careful and humble way.

Ove (non-indigenous) and Randi (non-indigenous) demonstrated the importance of the field work, real-world perception of real facts and realities about the challenges related to SD as well as the importance of opening and amplifying the mind-set when thinking about Sustainable Development.

5.2.3 *Juniper University (JU)*

At Juniper University (JU), the course encourages the students to look at the world in a non-reductionist manner.

EMMA (NON-INDIGENOUS): There is a tendency in the current education system in most Western developed nations to break down not only the silos of education but the reductionist view of the world.

Emma (non-indigenous) also emphasizes that the staff tries to foster a larger view of the world and a respect for those who do not have the same perspective or have a different position in relation to sustainability issues.

Felix (non-indigenous) and Rayan (non-indigenous) emphasized the importance of developing a critical mind-set and an intercultural viewpoint as capabilities derived from the course:

FELIX (NON-INDIGENOUS): The students learn to read literature, they learn to critique it, they learn to see it from an intercultural perspective and they also write essays, reports about it.
RAYAN (NON-INDIGENOUS): I hope that they draw out a critical mind-set around the institutions of society, both the socio-cultural institutions and legal and political.

5.2.4 Ice University (IU)

By considering the objectives, structure, pedagogies, and approaches of the course under study at Ice University (IU), the most important capabilities that could be pointed out are the understanding of the role energy and the ability to solve real-life problems. As Thomas and Noah state:

THOMAS (NON-INDIGENOUS): The biggest capability is to understand the role that energy plays in the whole picture of our lives. The basic thing is the student understanding the role energy plays in our lives. Hopefully, they also understand the role of fossil fuels in harming our life-support systems.
NOAH (NON-INDIGENOUS): The main capability is to be able to solve real life problems by using the energy science and knowledge of energy efficiency and renewable energy technology by creating options and analyzing the options to decide which the best option to solve a real problem is.

Apart from the analysis, evaluation of options and creation of real-life solutions, the interconnection of natural sciences and social sciences' concepts and the application of knowledge into practice is another important capability observed in the interview data. The focus is on sustainable energy and on how to apply energy science and knowledge of energy efficiency and renewable energy technology in real contextual problem-solving situations.

AUDREY (NON-INDIGENOUS): I think they (students) really start to piece together that information about natural sciences, physics and it gives them a very concrete information on how to apply the information.
OLIVIA (INDIGENOUS): They have a good idea about sustainable energy throughout the US and throughout the world.
NOAH (NON-INDIGENOUS): We see people without this knowledge using solar panels at home but using incandescent bulbs. After taking the course, they see they are willing to spend so many dollars on their project (and the objective is to reduce the consumption of fossil fuels); they realize that if they put the money (instead of using solar panels) on things like LED light bulbs or more insulation for the attic levels, they will achieve a much greater reduction on the consumption of fossil fuels. So, it is an example of how this knowledge can be used.

At the Ice University (IU) course, there seems to be an understanding that natural sciences, social sciences, and technology are completely interconnected and comprise interdisciplinary knowledge but, above all, the application of this knowledge contributes to solve real-life problems, to solve sustainability issues, to reduce the consumption of fossil fuels, and to save resources for the future.

References

Fish, D & Coles, C 2005, *Medical education developing curriculum for practice*, Open University Press, Maidenhead, p. 17.
Healey, M 2005, 'Linking research and teaching to benefit student learning', *Journal of Geography in Higher Education*, vol. 29, no. 2, pp. 183–201.
van der Horst, D, Harrison, C, Staddon, S & Wood, G 2016, 'Improving energy literacy through student-led fieldwork – at home', *Journal of Geography in Higher Education*, vol. 40, no. 1, pp. 67–76.

6 Curriculum design and review in the four case study institutions

6.1 Introduction

The curricula characteristics in the four case study institutions emerged from the documentary data and interview findings as per the convergences and divergences discussed in Chapter 5. In one of the case study institutions, energy is not exactly seen as a fundamental component of Sustainable Development Literacy (SDL). Currently, the curricula are more inclined to deal with energy within a natural resources management approach what does not give energy literacy, as an essential component of SDL, the application and the importance it has, to promote the vision of a sustainable energy system. The curricula do not seem to completely engage the stakeholders in their envisaged sustainable future in terms of the contents, approaches, and pedagogies. A sustainable energy system can be forged when everyone is equipped to operate on it. This reflects the importance of the curriculum design and review process as a dynamic instance to adapt and update components to engage and equip individuals towards a sustainable energy system.

There are specific processes for curriculum design and review adopted at Glacial University (GU), Borealis University (BU), and Ice University (IU), but not at Juniper University (JU). In this chapter the way curriculum is treated in the case study-universities will be described, as well as what could be improved in the energy curriculum according to staff interviews data.

6.2 Curriculum design and review

6.2.1 Staff and student meeting in the end of the term

At Glacial University (GU) there is a specific process adopted to design and review the curriculum, consisting of setting a common meeting with the students at the end of the course and before they leave the university to their home towns, in order to consult them on the elements that worked and the ones that didn't work along the course. This is an opportunity to hear suggestions for improvement, and in the latest curriculum review, for example, the course leader realized that one of the suggested improvements were related to the number of students that each mentor would supervise. Each mentor was assigned less than ten students (about eight students) to mentor in order to have a closer contact with them. The mentors are responsible for collecting suggestions from the students at the end of the course and an internal process of review happens with inputs of the lecturers, group coordinators, and course leaders resulting in a final and more formal report to document the process. Eva (indigenous) brought on board this report during the interview:

EVA (INDIGENOUS): It is a process in a few steps. First of all, we have a common meeting before the students leave, to ask them about how they felt along the course and how it had gone and what they liked and disliked, etc. This is in order to have a very good conversation about suggestions on how to improve. Of course, the students have a lot of work to do before they complete their exam, so since we have four mentors and we divide the students among each other, so each of us have less than ten students (about eight students) to mentor so we are able to keep very close contact with our students, so it is also the mentors' responsibility to hear from the students after their exams are completed about what we can do differently from their perspective but of course others than students have opinions, the lecturers, the group coordinators, also get some ideas on how we can do this differently and this review process is taken more internally among ourselves in a common discussion and finally, the results of the review process are written in our report; something more formal and to maintain the institutional memory of Glacial University.

GRETE (INDIGENOUS): The process consists in a meeting before the students leave the course to assess their opinion about what had worked and had not worked in relation to their expectations. There is an open conversation about suggestions on how to improve. Students, lecturers and coordinators give ideas on how we can do differently, and this review process is internal.

IVAR (INDIGENOUS): The students have to say something about how the course worked in comparison to the curriculum; how they had expected it and the learning outcomes; how difficult and what was covered and not covered, they have an input, a voice about it.

The curriculum review seems to be a reflective process at Glacial University (GU) and there is a current trend of adjusting the curriculum to adopt more than one perspective in the course plans. This trend reflects the content and the literature adopted. Kristine (indigenous) mentioned this current trend as follows:

KRISTINE (INDIGENOUS): You cannot really get a discussion going if you only take one view, you need more views, different views to have a reflection on any matter. I think you need more information, more knowledge and we have been working on influencing the curriculum to adopt more than one view, more than one perspective on the course plans. It means that we have a say on what books will be picked up, and what the subjects are in these books; this is how we want to influence or orient the curriculum.

At the Borealis University (BU), the curriculum review is a frequent process of curriculum adaptation regarding new contents, approaches, and pedagogies. Meetings with lecturers are set to discuss these elements and they run periodic evaluation schemes to stimulate students' participation to have opinions how to improve the curriculum. The process is not only a curriculum review but a manner to improve the relationship between lecturer and student, a way to engage the student a bit more into the subject and studies. Arne commented about this process:

ARNE (NON-INDIGENOUS): It is our interest to change and we meet the other lecturers and we find new contents, we do review quite often, you see how the lecture goes and you see if the students are able to absorb the information and you reconsider next time; we have the evaluation schemes and this is how to get participation for these evaluation schemes, I think that there is a challenge between the teacher-student culture something that needs to be developed here in Borealis University to help students to be more involved in lectures and research, to improve that relationship, to make it more personal. I think that the major barrier for that is how to structure the scheme, how to organize lectures and the timetable and this sort of research-teaching divide and these are sorts of challenges that are there and we are still developing the relationship between the lecturer and the students.

As Randi (non-indigenous) observes, the process has an involvement of students as there is a system where a student group is nominated to collect students' assessment of the course through a questionnaire or group discussion resulting in what needs to be adjusted to the following term. It happens in two occasions, as there is a midterm and end-term course evaluation and the suggestions are compiled into a final report:

RANDI (NON-INDIGENOUS): You know we have a system here through which you nominate a student group consisting in two or three people responsible for the student assessment of your course and then I discuss with them the methodology to use; if it is a big group we can do a questionnaire, in the learning platform or they can do a discussion with other students and then come to me with what can be improved so we use these tools and midterm course evaluation what can be done better, what is difficult, what is required, and we have a final assessment, when the final assessment is done we have to report it on the platform and this is for all courses. I present my assessment report where I address if there are critical remarks and I have to say how I improve it for next time.

6.2.2 Curriculum review via module evaluation

At Ice University (IU) there is a specific energy curriculum and the curriculum design and review are interrelated processes based on the lecturers' experience and student feedback. In general, in North America and Canada, course curricula are elaborated by the lecturers responsible for the subjects what gives the lecturers a high level of autonomy in the review process. However, in the particular case of Ice University course, the students' feedback is an important component of the curriculum review and there is an online midterm and final-term module evaluation regarding the lecturer, the content, the activities, assessments, and approaches. Noah is an energy expert and the course/programme leader in charge of these processes, and he observes:

NOAH (NON-INDIGENOUS): The process to review the energy curriculum, to develop this course in the first place, and to review it are processes that are related; the process to develop the course was to use my practical experience from the field and the knowledge I needed in the field to come up with solutions; to use those concepts and to shape them into a course it was my approach and this is the same approach I use to review it, and further revise the course. I propose the changes in the curriculum and then the curriculum review committees, at the university, review it as well, so this is the process.

An important point is the students' feedback that is incorporated in the curriculum and, according to the reports, it brings important benefits to shape and improve the curriculum.

OLIVIA (INDIGENOUS): The course leader developed all the materials but I review them eventually; he has created it and he always welcomes feedback; and it is recorded so I cannot really change it but he incorporates feedback from the students because I know that the feedback the students give in the beginning of the course helps to shape it and it is a big benefit doing that.

THOMAS (NON-INDIGENOUS): The process in this university is good but the board of leaders of the university probably does not focus on energy as they needed to. But our programme head started the programme and he is the person in charge of making changes.

In Juniper University (JU), as Emma observed in the previous chapter about the course structure, energy is implicit to students' lives. The result of that is the inexistence of an energy curriculum.

EMMA (NON-INDIGENOUS): As I said, energy is implicit, we don't have a specific energy curriculum.

The difference identified in the curriculum making and review at Juniper University (JU) is that there is not much review as per Felix reports:

FELIX (NON-INDIGENOUS): At the Juniper University, I don't think there really is a lot of curriculum review; they use programmes and they decide to cut courses or add courses, but in north America we have a high degree of autonomy and people teach pretty much what they want to teach, it is all mediated through fairly detailed course syllabus, resources and course outlines.

Rayan (non-indigenous) confirmed the previous position by stating that there is no rigorous curriculum review or institutional process, and Zach (non-indigenous) observed that the demand or undersubscribed courses influence the course options to be offered as follows:

RAYAN (NON-INDIGENOUS): To review the curriculum we don't have a rigorous curriculum review in our institution like you have in some American institutions. There are some benefits and costs of it, the benefits are that we really have full academic freedom, nobody really cares of what I teach in my courses, I am freely do what I want. I can use a textbook, I can bring what I think is particularly important what means I can bring my own perspective and what I think makes my course stronger. The reviews are on my own work and the students evaluate me I take on feedback and I adapt it and make changes; but we don't have a third part review even collectively as a group; we don't spend a lot of time looking at each other's courses, we chat about what each lecturer will cover; the only review is that I am supposed to provide the course objectives and outline, but nobody reviews it.
ZACH (NON-INDIGENOUS): Within the energy unit we will add courses as we see a demand for them and people pitch them or if the courses are undersubscribed, other options will take them out of the streamline or change the content.

6.3 Curriculum improvement

In terms of curriculum improvement, Eva (indigenous), from Glacial University (GU), mentioned a relevant point on how to balance the local and the global levels and how to communicate global and local models in a more efficient way as follows:

EVA (INDIGENOUS): we should, perhaps, focus on the post aspects of the course, on how much we have managed to communicate these types of global models and the local relevance (...) this is something that I feel we could improve.

EVA (INDIGENOUS): A lot of local communities are facing land-use battles, so the local level is very important but on the other hand we have seen here in the Arctic that the local value is very important, but the global challenges comes to us very quickly. Global climate change affects us first here in the Arctic; perhaps the global is more relevant to us here in terms of environmental models.

Eva (indigenous) also mentioned that she wishes to design a more creative curriculum to reach a better understanding in relation to the global models, i.e. climate change applied to a local context, as it seems students have this need to contextualize the global patterns to the local context. Concerning this point, she says:

EVA (INDIGENOUS): The global and the local? We need a more creative curriculum, to do more research and making the global models more relevant at local level. (...) the students perceive it a lot more differently when you tell them to 'translate a global climate change model' to the town context. I believe that it is a good way of creating interest.

Ivar (indigenous) also pointed out that the curriculum could approach more technical issues; it could be more practical and improve even more the application of theories.

IVAR (INDIGENOUS): Energy use, or energy challenges or oil and hydrocarbons and radioactivity and it could be a little more practical. It could be more technical, how to build windmills, etc. But you need to have this connection about the practice and the theories.

As reported in Chapter 5 (see Section 5.1.1.4B) related to the curriculum structure, Nils, from Borealis University (BU), is of the opinion that the curriculum needs change in terms of the mindset. Nils was a lecturer who described the important mission of the geographer in terms of promoting Sustainable Development (SD). According to him, SD is studied more in ecology than in geography and he also expresses his concern to the fragmented (compartmented), disintegrated/unconnected, and not embedded view of SD in different topics studied and reflected in the course content as follows:

NILS (NON-INDIGENOUS): People should change their mentality about re-managing the curriculum.

NILS (NON-INDIGENOUS): the curriculum must show that Sustainable Development is not something about ecology, ecosystems or environment, this is something about economy, sociology about lifestyle about all these things; you have to show this; we need the geographer not separated from the physical geographer, from the climatologist from the ecologist.

NILS (NON-INDIGENOUS): but it is separated; here the problem is that; we need a curriculum with a complete paradigm;

NILS (NON-INDIGENOUS): We don't have enough literature of Sustainable Development, not enough books with the views about Sustainable Development. All the books about Sustainable Development are written by ecologists or economists but are not written by people that know the relationship between Sustainable Development and sociology or can make relations between different aspects of Sustainable Development.

Improvements to the curriculum were suggested by Nils (non-indigenous) and Randi (non-indigenous). Nils believes that new literature needs to be introduced to the curriculum.

NILS (NON-INDIGENOUS): (...) we need new literature, we need new programmes and new views involved in this issue of Sustainable Development, we need new literature, specially.

Randi (non-indigenous) defends the application of what is studied in the course on other geographical areas and she believes in a more proactive curriculum that emphasizes the application and the capabilities acquired in the course. She says:

RANDI (NON-INDIGENOUS): I think that Sustainable Development must be taken home; to the home country of several students because they think about Sustainable Development in different ways.

She also thinks that the curriculum could be improved to make the students to reflect the harsh reality of our lives and to reflect educators' responsibility of disseminating the issues related to SD.

RANDI (NON-INDIGENOUS): it is very easy to forget what you learnt at the university, it is more treated like an idea; they are not reflecting the harsh reality of our lives. I am preaching about the responsibility we have of disseminating these issues of Sustainable Development and capabilities. These are the values and the norms we must pursue to improve the curriculum.

For some lecturers at Ice University (IU), the curriculum could be improved by presenting technical content in shorter segments to attend the needs of indigenous students that present a different way of learning, but mainly by continuously updating the content of renewable energy technologies and the communication technology, as the course has an in-class and

160   *Curriculum design and review*

online component to attend students across the Arctic. Noah (non-indigenous) commented about possible improvements:

NOAH (NON-INDIGENOUS): the first thing is to update the content with fresh newest technologies mainly because renewable energy is a quickly evolving field and if I want to provide information that is useful, I need to provide updated information.

Olivia (indigenous) also observed that contextualization would make the content more engaging:

OLIVIA (INDIGENOUS): for this course, I think that the communication technology is something that could be improved, but for the content itself, the information should be presented in a more engaging manner; having a little more real-world examples, more pictures from the villages would make it a bit more engaging and relevant; by using information and examples that are specific to students' culture and location.

In terms of curriculum improvement, Rayan, from Juniper University (JU), called attention to the need for a more integrated approach to understand the connections of sustainability (energy inclusive), as students can combine modules but, in the end, they don't have an integrated view of them. Rayan (non-indigenous) says that:

RAYAN (NON-INDIGENOUS): I think we really fail in providing students an education in the breath of approaches to sustainability as a whole, so some people will take a course with an ecologist and will get a learning about limits to growth and about the challenges we have at that front, other people will take a course with another lecturer and learn about innovation and how we drive change; other people will take a course in sociology with individuals on campus who very much see the world as if been driven by capitalist testimony. I think we need to help students to navigate this breath because they go from course to course and it took over a decade to really understand the connection and integration of these different areas and we take all of these students and we see there is a debate, this person is right, and this person is wrong or this is the right approach and this is the wrong approach. I think we must help them now to understand the integration of these different areas of knowledge; the really important connection between climate change and energy.

On the other hand, Zach (non-indigenous) observed that no improvement is required, as the programme is shaped around skills not based on a subject area and energy can mean different things to different people:

ZACH (NON-INDIGENOUS): I don't think it could be improved. I tend to think that energy is better as a minor not as a major. I would say you should do accounting with an energy minor, or finance with an energy minor, because energy can mean so many things for so many different people, so we don't tend to structure programmes around a subject area we tend to structure them around a skill set.

In sum, this chapter raised important differences how curricula are designed and reviewed in different cultural environments. The curriculum design and review are interconnected processes influenced by the local economic and cultural mindset varying from a more participative experience of co-creation to a more pro-forma and personal exercise.

7 Levels of energy literacy and adherence to ESD
Arctic Citizenship Model

7.1 Introduction

This chapter aims at presenting the documentary findings to allow an evaluation of embedded levels of energy literacy and the course structure alignment and adherence to ESD. It also outlines and co-relates the document findings on levels of energy literacy and ESD adherence to (staff and students) interview findings of the case studies. The findings contained in this chapter inform the discussion and analysis developed in Chapter 10. These facets are contrasted based on documents, and perspectives of lecturers and students that operate in Arctic Norway, Canada, and Alaska. The process of documentary analysis involved skimming, and then a comprehensive reading and interpretation combining content analysis and thematic analysis, by organizing information into categories related to the central questions of the study (Bowen 2009, p. 32) and searching for pattern recognition (via NVivo 12) within the data, with emerging themes becoming the categories for analysis based on a consistent set of documents in order to make a systematic comparison on all learning outcomes in programme handbooks, syllabuses, identifying courses content, skills and competences being developed, and compulsory assignments.

7.2 Levels of energy literacy

Chapter 2 highlighted the importance of energy as a geographical concept and energy transition as a geographical process, involving the reconfiguration of current patterns and scales of economic and social activity according to Bridge et al. (2013) in his paper 'Geographies of energy transition: Space, place and the low-carbon economy', setting an important premise of this research.

At that occasion, energy studies and human geography/sustainability programmes were identified as an opportunity to critically examine the process of energy transition. Along this path, it was identified that HE has a much more relevant role than only acknowledging this process of transition. HE

Table 7.1 Arctic Pedagogical Model of Energy Literacy based on cognitive, affective, behavioural and capability components, and findings. The italics represent the findings from this study that are new, whereas citations indicate those drawn from the existing literature

Arctic Energy Literacy subscales	Pedagogical foundation/typology	Arctic Pedagogical Model of Energy Literacy
Cognitive	Skill and knowledge-based	To have *conceptual knowledge* about energy
Cognitive	Skill and knowledge-based	To have meaningful understanding of the *everyday life energy use*
Cognitive	Skill and knowledge-based	To have meaningful understanding of the *impact* of energy production/use on *environment and society*
Cognitive	Skill and knowledge-based	*To have meaningful understanding of the impact of personal educated energy decisions and actions on global community*
Cognitive and affective	Attitudes and values-based	To have meaningful understanding and empathy of the needs for: • *Energy conservation* • *Alternatives* energy resources to fossil fuels
Capability	Critical thinking informing educated energy decisions	To have the capacity to assimilate and critique information to inform choices, attitudes and sustainable behaviours (St. Clair 2003).
Capability	Transformative learning-based	To reflect upon problematic perspectives and mind-sets (Mezirow 2009) = *to reflect on complexity = complexity thinking capability*
Capability	Cultural capability-based	To deal with the 'other' *cultural capability for Sustainable Development*
Capability	Technological capability-based	To deal with technology
Affective and Behavioural	Attitudes, values and behaviour behaviour-based	To make choices and exhibit behaviours reflecting attitudes according to this understanding concerning energy resource development and consumption = *well-informed decision-making*
Affective, Behavioural	Applications to real life (constructive learning) problem-based/learning from direct experience/ active learning Local-global context	Where a route from current knowledge to new information is paved through applications to real life (Hein 1991) *(innovation)* *real-world settings* make the authenticity of learning activities clear and connect the lecture theatre to the home. Learning actively changes the way an environment is viewed and interacted with (Kolb 1984).

presents the potential to insert students into this reality in practical terms. This is the central concern of this section: to understand how HE is doing that. The relationship between energy and geography is intertwined and highly complex. As van der Horst et al. (2016) observes, cheap and abundant energy determines the type of geographical constraints that influences societies to deal with space (land), climate, sustainability, and other societies (globalization).

In Chapters 2 and 3, references were made to studies carried out by a few scholars in relation to the importance of improving energy literacy in HE not only to enable people to exercise ESD but to be able to make educated energy decisions to contribute to SD. van der Horst et al. (2016) for instance, pointed out the importance of and the need to improve energy literacy, equipping people to make more thoughtful, responsible energy-related decisions and actions.

There are, in fact, a few studies covering this topic and there is no definite or final model to assess energy literacy levels in HE courses in the Arctic or non-Arctic nations. There are important conceptualizations of energy literacy comprising cognitive (knowledge), affective (attitudes and values), and behavioural skills but the pre-existent theoretical framework seems insufficient to measure levels of energy literacy in HE.

Despite the model, presented in Chapter 3 (Section 3.4.1), seeming quite expanded, the empirical information acquired in the Arctic field work (primary data) shows that there are other fundamentals to be considered for a comprehensive Arctic Pedagogical Energy Literacy Model related to teaching energy geographies (see Table 7.1).

To measure baseline levels of energy literacy and assess broader impacts of educational interventions, it is important to go beyond cognitive (content knowledge, cognitive skills), affective (attitudes, values), and behavioural (including predispositions to behave) components. Additional components such as individual and collective decision-making in real-world contingences, intergenerational impacts knowledge, and cultural capabilities appeared as relevant aspects to be literate in energy in the Arctic. Improving students' energy literacy will empower them to become engaged in, objectively, assessing energy-related decisions throughout their daily lives. Moreover, it will empower learners to create and participate in real-world solutions contributing to the welfare of their communities realizing the core objective of an effective and informed global citizenry through geo-capabilities (Walkington et al. 2018).

A more comprehensive model for assessing energy literacy (see Table 7.2 and Figure 7.1) is proposed in this book by adding up more practical components from the primary data, as follows:

Table 7.2 Innovative Arctic Pedagogical Model of Energy Literacy showing the relevant educational components present in the courses under analysis based on curricular objectives

Arctic Energy Literacy subscales	Pedagogical foundation/typology/	Arctic Pedagogical Model of Energy Literacy	GU curricular objectives	BU curricular objectives	JU curricular objectives	IU curricular objectives
1. Cognitive	Skill and knowledge-based	To have conceptual knowledge about Sustainable Development and energy (natural and social sciences)	Basic academic approaches/frameworks to biological diversity (1)	Combines natural sciences, humanities and social sciences development process in developing countries and the problems related to climate change Globalisation and Sustainable Development Natural Resources Management (1)	Science, policy and social processes natural resources, energy and environmental issues and industries Science, policy and social processes (1)	Yes – science concepts related to energy economic aspects of sustainable energy societal problems and solutions related to its energy use and production Sustainability of current energy practices (1)
2. Cognitive	Skill and knowledge-based	To have meaningful understanding of the energy use in everyday life	n/a (0)	n/a (0)	n/a (0)	Analysis of energy systems (1)
3. Cognitive	Skill and knowledge-based	To have meaningful understanding of the impact of energy production/use on environment and society	Indigenous economic activities impacted by energy production and use (1)	Policies and economy of renewable energy and fossil fuels (1)	Community, health & environmental issues, environmental hazards (1)	Analysis of energy systems and practices (1)

(Continued)

Table 7.2 (Cont.)

Arctic Energy Literacy subscales	Pedagogical foundation/typology/	Arctic Pedagogical Model of Energy Literacy	GU curricular objectives	BU curricular objectives	JU curricular objectives	IU curricular objectives
4. Cognitive	Skill and knowledge-based local-global context	To have meaningful understanding of the impact of personal educated energy decisions and actions on local and global community Application at home but also application to the world.	Use knowledge in practice – knowledge dissemination in communities (1)	n/a (0)	n/a (0)	Practices of energy production and use (1)
5. Cognitive and affective	Attitudes and values-based	To have meaningful understanding and empathy of the needs for: -Energy conservation -Alternatives energy resources to fossil fuels = Sustainable energy	Systematic and ethical manner of using natural resources (1)	Climate, energy- and energy policy (1)	n/a Environmental issues, including renewable and non-renewable natural resources. (0)	Energy efficiency and renewable energy approaches, sustainable energy (1)
6. Capability	Critical thinking informing educated energy decisions	To have the capacity to assimilate and critique qualitative and quantitative information to inform energy choices, attitudes and sustainable behaviours (adapted from St. Clair 2003).	Reflect on his/her own empirical work (1)	n/a (0)	n/a environmental management (0)	To analyse and evaluate the feasibility of possible solutions to energy problems the economics of sustainable energy (1)
7. Capability	Transformative learning-based	To reflect upon problematic perspectives and	n/a (0)		n/a (0)	

		mind-sets (Mezirow 2009) = to reflect on complexity = complexity thinking capability		Problems at local, regional and global levels (1)		Problems with current energy practices; solutions to current energy problems (1)
8. Capability	Cultural capability-based	To deal with the 'other' – anthropological view Cultural capability for Sustainable Development	Core academic approaches and frameworks to Traditional Knowledge (1)	n/a or optional component for the students that choose to have indigenous studies (0)	n/a or optional component for the students that choose to have indigenous studies (0)	n/a (0)
9. Capability	Individual and collective decision-making	To produce individual and collective impact	Assessment field work, to be presentation, dissemination (1)	Public participation – impact (1)	n/a (0)	Societal problems and solutions related to its energy use and production (1)
10. Capability	Technological capability-based	To understand the technological factor of energy	n/a (0)	New renewable energy technologies (1)	n/a (0)	Literature and practice regarding energy technologies (1)
11. Capability	Intergenerational	Intergenerational dimension	Innate of the cultural group	Intergenerational impacts	n/a (0)	Intergenerational impacts

(Continued)

Table 7.2 (Cont.)

Arctic Energy Literacy subscales	Pedagogical foundation/typology/	Arctic Pedagogical Model of Energy Literacy	GU curricular objectives	BU curricular objectives	JU curricular objectives	IU curricular objectives
	Impacts of energy production and use		Cultural characteristic of indigenous mind-set (1)	(1)		(1)
12. Capability	Research	Review literature, research and compile information	Knowledge of methodologies for documenting Traditional Knowledge Plan and conduct research according to methodology; Multidisciplinary research (1)	Quantitative and qualitative methods Interdisciplinary orientation (1)	Research Methods in Human Geography and Planning (1)	Compile information related to sustainable energy (1)
13. Affective, Behavioural	Attitudes, values and behaviour behaviour-based	To make choices and exhibit behaviours reflecting attitudes according to this understanding concerning	n/a (0)	n/a (0)	n/a (0)	Apply the knowledge to real-life energy issues (1)

14. Affective, Behavioural	Applications to real life (constructive learning) problem-based/learning active learning Local-global context	energy resource development and consumption= critical thinking decision-making Where a route from current knowledge to new information is paved through applications to real life (Hein 1991); innovation and application to solve real world problems real-world settings makes the authenticity of learning activities clear and connects the lecture theatre to the home. Learning actively changes the way an environment is viewed and interacted with (Kolb 1984).	Experience-based knowledge field-projects Exchange views and shared experiences (1)	Problems at a local level, as well as to regional and global levels field work projects (1)	n/a (0)	Solutions to current energy problems Apply the knowledge to real-life energy issues (1)

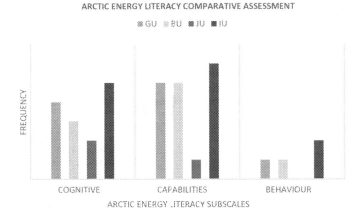

Figure 7.1 Analytical result based on the Innovative Arctic Pedagogical Model of Energy Literacy applied to case studies. Comparative assessment produced based on the presence or absence (frequency) of the course components from subscales 1 to 14 shown in Table 7.2.

7.2.1 Glacial University (GU)

The course at Glacial University (GU) is structured to have a place-based component and an online component. This dual-component structure benefits populations across the vast distances of the circumpolar Arctic as the intent is to reach a multilingual indigenous population in towns and the most remote communities. It has a circumpolar perspective focused on Arctic biodiversity according to the Convention of Biological Diversity elements. The course is contextual, interdisciplinary and focused on environmental, societal, and multicultural aspects approached through indigenous perspectives and how to document these perspectives through empirical research. Energy is a component but not the emphasis. Intergenerational and cultural aspects are emphasised.

This course is administered as a bachelor's degree taught in English and in other languages (indigenous and non-indigenous). It has a focus on connecting analytical and empirical approaches to TK. Academic debates on how TK contributes to sustaining indigenous peoples' societies and the role of TK for conservation of biological diversity are stimulated. The course provides students with practical experience of theory application by using methods to document TK on biological diversity in a systematic manner, by using real-life settings (home communities) in order to engage the students into the study of their own realities and to stimulate knowledge application in real-world circumstances impacting local peoples. At the same time, it allows them to apply their knowledge into the studied site, it also helps them disseminate the learning experience to a vaster spectrum as sharing experiences with the group (staff and students) contributes to spread the learning to a circumpolar scale.

7.2.2 Borealis University (BU)

The course at Borealis University (BU) combines natural sciences, humanities, and social sciences with emphasis on developmental process in developing countries and the problems related to climate change and natural resources management. Interdisciplinary aspects related to SD and globalization are emphasized at the level of economy and policies of renewable energy and fossil fuels. The case studies (problems) examined relate to local, regional, and global geographical areas where fieldwork projects are developed through quantitative and qualitative research methods under interdisciplinary orientation. Energy is a component of natural resources management.

7.2.3 Juniper University (JU)

Science, policy, and social processes are studied with an emphasis on natural resources, energy and environmental issues, and industries. The student can combine courses based on his/her interests. Components are related to economics, policies, cultural studies (sociology), and sustainability. Energy is an implicit component and more emphasis is given to health and environmental issues, environmental hazards, and impacts.

7.2.4 Ice University (IU)

Energy literacy is addressed and emphasised directly through the course at Ice University (IU). Science concepts related to energy, sustainable energy, energy technology, economic aspects of sustainable energy, societal problems, and solutions related to its energy use and production are clearly examined and addressed.

Students are exposed to analysis of current energy practices, sustainability assessment of these practices, and they are expected to produce analysis of energy systems and practices in relation to energy production and use. The course stimulates students to know about energy efficiency and renewable energy approaches and to evaluate the feasibility of possible solutions to energy problems in real-world case studies at local level and create solutions to current energy problems. The economics of sustainable energy is also addressed, and the course aims at reaching every person regardless of age and profession. The mind-set is that energy is important to everybody to make decisions in everyday life.

7.3 Adherence to ESD

Vare and Scott's framework (Table 3.5 [Two Sides of ESD] and Table 3.6 [ESD1 and ESD2]) is useful and contains relevant components to establish stages in which ESD is present in the curriculum. However, this framework seems insufficient facing the complexity of elements of this research data collection. By transposing elements from the findings in Table 3.6, new elements of the framework are proposed in Table 7.3:

Table 7.3 ESD1 and ESD2 positive and negative aspects based on Vare and Scott (2007, pp. 3, 4), including findings in italics

ESD 1	ESD 2
Positives	**Positives**
• Promotion of informed, skilled behaviours and ways of thinking	• Building capacity to think critically about what experts say and to test ideas *(critical thinking as an important capability)*
• Useful in the short-term where the need for this is clearly identified and agreed *(useful as a rhetoric)*	• Exploring the dilemmas and contradictions inherent in sustainable living *(complexities)*
	• A necessary complement to ESD 1, making it meaningful in a learning sense *(meaningful concepts and learning)*
Negatives	
• People rarely change their behaviour in response to a rational call to do so *(to talk about values is not enough to change behaviour)*	
• Too much successful ESD 1 in isolation would reduce our capacity to manage change ourselves and therefore make us less sustainable *(static)*	
• Not adequate for the long-term *(not adequate to the intergenerational approach)*	

Table 3.7 (Three Types of ESD – Sustainable Development, learning, and change, Scott and Gough (2003, pp. 113, 116) – is also a useful framework to understand the opportunities and challenges of the possible types of ESD, what makes the case for building up a more comprehensive model as per Table 7.6 to assess levels of ESD in the four case study universities chosen

Levels of energy literacy and adherence to ESD 173

for this research. According to Scott and Gough (2003, pp. 113, 116), there is 'learning about SD', 'learning for SD', and 'learning as SD', which leads to use of the three-type model to define levels of adherence to ESD. These are the three basic levels to start an assessment of the adherence of curricula to ESD. The three levels represent a gradation from a more passive way of learning (a less active and less participative) towards a more active and creative application of the knowledge acquired in HE.

In order to be able to innovate the pedagogical systems of teaching SD and energy, it is important to set the framework with the core elements to assess levels of adherence of curricula to ESD. The emergent model based on the proposed framework is useful and contains relevant components to establish stages in which ESD is present to the curricula; however the data collected and presented in previous chapters (findings) of this book could serve to add up some new elements fundamental to design a new and more comprehensive framework to measure levels of adherence to ESD. Based on the three approaches of Scott and Gough (2003, pp. 113, 116) and developing the model by adding elements from the data collection (interviews and documents), the fundamental components for Arctic education are proposed in Table 7.4:

Table 7.4 ESD adherence model based on Scott and Gough (2003, pp. 113, 116). Three types of ESD and notes related to Arctic data collection in the four case study universities and curricula. The words in italics and bold represent the elements from the findings acquired during the Arctic research

Type 1 – Learning about SD *(Passive)* Sustainable Development as being expert-knowledge-driven where the role of the non-expert is to do as guided	**Type 2 – Learning for SD** *(Action is desired)* Sustainable Development as being expert-knowledge-driven where the role of the non-expert is to do as guided	**Type 3 – Learning as SD** *(Action)* Expert and learner work together for a tangible outcome (impact)
• Assume that the problems humanity faces are essentially environmental *(it is just about management)*	• Assume that our fundamental problems are social and/or political, and that these problems produce environmental symptoms *(it is about economy, environment, society)* **environmental impacts**	• Assume that what is (and can be) known in the present is not adequate

(*Continued*)

Table 7.4 (Cont.)

• Can be understood through science and resolved by appropriate environmental and/or social actions and technologies *(the hope is technology to resolve problems)* • It is assumed that learning leads to change once facts have been established and people are told what they are *(passivity)*	• Such fundamental problems can be understood by means of anything from social-scientific analysis to an appeal to indigenous knowledge • The solution in each case is to bring about social change *(activity is desired)* • Learning is a tool to facilitate choice between alternative futures which can be specified on the basis of what is known in the present *(learning can provide choices and decisions)*	• Desired 'end-states' cannot be specified • This means that any learning must be open-ended • Essential if the uncertainties and complexities inherent in how we live now are to lead **to reflective social learning** about how we might live in the future

We go beyond the learning as participation to propose learning as an 'active co-creator of adaptive solutions to complex and unpredictable real-world problems', a model that reveals the 'Arctic Citizenship Model' or the 'Education for Arctic Citizenship' as shown in Table 7.5 – ESD adapted framework from Scott and Gough (2003, pp. 113, 116), Haigh (2014, p. 14), Walkington et al. (2018, p. 11) – by adding the findings from Arctic data collection.

Education for Arctic Citizenship goes beyond 'learning as participation' (Vare 2007) this is learning as sustainability in action. It is learning that is inculcated in the students' minds as per their practical (active) familiar, geo-experience of creativity and dissemination when living together. It is adaptive open-minded knowledge whose purpose is to effectively impact at local, regional, and global arenas. This is the 'education to be a geo-citizen', where the education site respects the 'space' and 'time' of the geographical area at stake, the Arctic.

The findings (documents and interviews) demonstrated that the knowledge and skills required to live and interact in the Arctic require more

Table 7.5 Arctic Citizenship Model. Five Types of ESD adapted framework from Scott and Gough (2003, pp. 113–116) and Haigh (2014) by adding up the findings from Arctic data collection

Type 1 – Learning about SD (Passive)	Type 2 – Learning for SD (Action is desired)	Type 3 – Learning as SD (Adaptive learning)	Type 4 – Education for Arctic Citizenship	Type 5 – Education for Global Citizenship
Sustainable Development as being expert-knowledge-driven where the role of the non-expert is to do as guided	Sustainable Development as being expert-knowledge-driven where the role of the non-expert is to do as guided	Expert and learner work together for a tangible outcome (impact)	Multicultural experts and learners are co-creators and co-managers of knowledge Arctic Knowledge is the fruit of the encounter of documented TK and western knowledge Arctic Knowledge includes different perspectives (indigenous and non-indigenous perspectives) The Arctic citizen demonstrate geo-capabilities: • Geographical imagination • Ethical subject-hood • Integrative thinking about society and environment • Spatial thinking • A structured exploration of place The curriculum presents multicultural perspectives and foundations	Self-identified with the whole of humanity not only tribal or group family Knowledge for living together sustainably, peacefully and responsibly
Assume that the problems humanity faces are essentially environmental	Assume that our fundamental problems are social and/or political, and that these problems	Assume that what is (and can be) known in the present is not adequate;	The Arctic citizen is eco-literate, energy literate and TK literate The knowledge is active, dynamic, in constant adaptation and it is used as an adaptation tool and	The global citizen is eco-literate (energy literate) and

(Continued)

Table 7.5 (Cont.)

Type 1 – Learning about SD (Passive)	Type 2 – Learning for SD (Action is desired)	Type 3 – Learning as SD (Adaptive learning)	Type 4 – Education for Arctic Citizenship	Type 5 – Education for Global Citizenship
(it is just about management)	produce environmental symptoms *(it is about economy, environment, society) environmental impacts*		empowerment for indigenous and non-indigenous groups – no ethnocentric *(adaptability to complexity/dynamic systems) the SD is continuously in the making like the ESD model is in constant adaptation because knowledge is a dynamic system.*	interacts constructively across cultural boundaries
Can be understood through science and resolved by appropriate environmental and/or social Actions and technologies *(the hope is technology to resolve problems)*	Such fundamental problems can be understood by means of anything from social-scientific analysis to an appeal to indigenous knowledge.	Desired 'end-states' cannot be specified	Arctic Knowledge is opened and constantly passible of being widen *(adaptability to unpredictability as knowledge is not static and can be widen by looking at the 'other' to different types of knowledge (TK) (technology and social-technology) (as systems are dynamic, learning is also dynamic) (activity-active learning and application) can create new realities.*	Knowledge oriented to the interconnectedness of different geographical areas Developing understanding, emotional intelligence (empathy and compassion)
It is assumed that learning leads to change once facts have been established and people are told what they are. *(passivity)*	The solution in each case is to bring about social change *(activity and application are desired)*	This means that any learning must be open-ended.	knowledge creates solutions and new realities for the Arctic citizens that have a vision of the 'Arctic nation' and the Arctic citizen indistinctively of ethnicity Knowledge to live together sustainably in the Arctic	Knowledge that embeds justice, fairness, responsibility

Learning is a tool to facilitate choice between alternative futures which can be specified on the basis of what is known in the present. *(learning can provide choices and decisions)*	Essential if the uncertainties and complexities inherent in how we live now are to lead to reflective social learning about how we might live in the future.	No pedagogic imperialism New way of being (authenticity and identity) being indigenous or non-indigenous, having the right to be and express yourself. *(individual and collective reflectivity- individual and collective critical thinking)* critical reflection becomes a collective exercise; critical reflection is a social and multicultural process. Critical reflection is a multicultural process) *Social critical thinking as learning in social action*	No pedagogic imperialism New way of being in the world – more inclusive
		Leading to real-life learning experience	
		Creating solutions for real and complex problems Making daily life well-informed and responsible decisions based on eco-literacy and energy literacy Disseminating good practices Impacting effectively the area (local, regional, global) In the long term- intergenerational outcome – future for all For a sustainable way of life/durable model of existence and wellbeing for all	Creative co-existence of nature and humanity or planetary citizenship (Henderson & Ikeda 2004)

than the ESD types 1, 2, and 3 from Scott and Gough (2003, pp. 113, 116). It requires a proper 'Education for Arctic Citizenship' that is not the traditional western education conveyed in an educational site. The concepts of space and time are variable as the perspectives of the stakeholders.

After framing a possible set of components/parameters to evaluate the level of ESD practiced at the locations under study, we proceed to the evaluation of the curricula adherence, as follows:

Table 7.6 ESD Adherence Parameters Analytical Model categorizing five types of ESD applied to the four case studies

ESD Adherence parameters	GU	BU	JU	IU
Type 1 – Learning about SD *(Passive learning)*				
1. Sustainable Development as being expert-knowledge-driven where the role of the non-expert is to do as guided	x	γ	γ	x
2. Assume that the problems humanity faces are essentially environmental (it is just about management)	x	x	γ	x
3. Can be understood through science and resolved by appropriate environmental and/or social actions and technologies *(the hope is technology to resolve problems)*	x	x	γ	x
4. It is assumed that learning leads to change once facts have been established and people are told what they are (passivity)	x	x	γ	x
Type 2 – Learning for SD *(Application/ Action is desired in learning)*				
5. Assume that our fundamental problems are social and/or political, and that these problems produce environmental symptoms (it is about economy, environment, society) environmental impacts	γ	γ	x	γ
6. Such fundamental problems can be understood by means of anything from social-scientific analysis to an appeal to indigenous knowledge	γ	x	x	γ
7. The solution in each case is to bring about social change (activity is desired)	γ	γ	x	γ

(*Continued*)

Table 7.6 (Cont.)

ESD Adherence parameters	GU	BU	JU	IU
8. Learning is a tool to facilitate choice between alternative futures which can be specified on the basis of what is known in the present (learning can provide choices and decisions)	γ	γ	x	γ
Type 3 – Learning as SD *(Adaptive learning)*				
9. Expert and learner work together for a tangible outcome (impact)	γ	x	x	γ
10. Assume that what is (and can be) known in the present is not adequate;	γ	x	x	γ
11. Desired 'end-states' cannot be specified	γ	x	x	γ
12. This means that any learning must be open-ended	γ	x	x	γ
Essential if the uncertainties and complexities inherent in how we live now are to lead to reflective social learning about how we might live in the future	γ	x	x	γ
Type 4 – Education for Arctic Citizenship				
13. Multicultural experts and learners are co-creators and co-managers of knowledge	γ	x	x	γ
14. Arctic knowledge is the fruit of the encounter of documented TK and western knowledge	γ	x	x	x
15. Arctic knowledge includes different perspectives (indigenous and non-indigenous perspectives) multicultural perspectives	x	x	x	x
16. The Arctic citizen demonstrates geo-capabilities: • Geographical imagination • Ethical subject-hood • Integrative thinking about society and environment • Spatial thinking • A structured exploration of place	γ (presenting all the elements)	x	x	γ (presenting all the elements)
17. The curriculum presents multicultural perspectives and foundations	x	x	x	x
18. The Arctic citizen is eco-literate, energy literate and TK literate	x	x	x	x
19. The knowledge is active, dynamic, in constant adaptation and it is used as an	x	x	x	γ

(Continued)

Table 7.6 (Cont.)

ESD Adherence parameters	GU	BU	JU	IU
adaptation tool and empowerment for indigenous and non-indigenous groups – no ethnocentric				
20. Arctic knowledge is opened to be constantly enhanced and expanded	γ	x	x	γ
21. Adaptability to unpredictability as knowledge is not static and can be widen by looking at the 'other' to different types of knowledge (TK) (technology and social-technology)	γ	x	x	γ
22. As systems are dynamic, learning is also dynamic; activity (active learning and application) can create new realities	γ	x	x	γ
23. Knowledge creates solutions and new realities for the Arctic citizens that have a vision of the 'Arctic nation' and the Arctic citizen indistinctively of ethnicity	x	x	x	x
24. Knowledge to live together sustainably in the Artic	γ	x	x	γ
25. No pedagogic imperialism	x	x	x	x
26. New ways of being (authenticity and identity) being indigenous or non-indigenous, having the right to be and express the self	γ	x	x	γ
27. (individual and collective reflectivity-individual and collective critical thinking) critical reflection becomes a collective exercise; critical reflection is a social and multicultural process. Critical reflection is a multicultural process) Social critical thinking as learning in social action	γ	x	x	γ
28. Leading to real-life learning experience	γ	x	x	γ
29. Creating solutions for real and complex problems	x	x	x	γ
30. Making everyday life well-informed and responsible decisions based on eco-literacy and energy literacy	γ	x	x	γ
31. Disseminating good practices	γ	x	x	γ
32. Impacting effectively the area (local, regional, global)	γ	γ	x	γ

(Continued)

Table 7.6 (Cont.)

ESD Adherence parameters	GU	BU	JU	IU
33. In the long term (intergenerational outcome – future for all)	γ	γ	x	γ
34. For a sustainable way of life/durable model of existence and wellbeing for all	γ	x	x	γ
Type 5 – Education for Global Citizenship				
35. Self-identified with the whole of humanity not only with a group or family	x	x	x	γ
36. Knowledge for living together sustainably, peacefully and responsibly	x	x	x	x
37. The global citizen is eco-literate *(energy literate)* and interacts constructively across cultural boundaries	x	x	x	γ
38. *The Global citizen demonstrates geo-capabilities*	γ	x	x	γ
39. *Knowledge oriented to the interconnectedness of different geographical areas*	x	x	x	x
40. Developing understanding, emotional intelligence (empathy and compassion)	γ	x	x	γ
41. Knowledge that embeds justice, fairness, responsibility	γ	x	x	γ
42. No pedagogic imperialism	x	x	x	x
43. New way of being in the world – more inclusive	x	x	x	x
44. Creative co-existence of nature and humanity or planetary citizenship (Henderson & Ikeda 2004)	x	x	x	x

Based on the ESD Adherence parameters proposed by the Analytical model originated by the data, it was possible to compare the level of adherence of the courses in the four case study universities as per Table 7.6 by considering the presence and the absence of the parameters associated to the five types of ESD, with special attention to the features of Type 4 identified as 'Education for Arctic Citizenship'. The components identified in the courses from the four case study universities provided scores used to

182 Levels of energy literacy and adherence to ESD

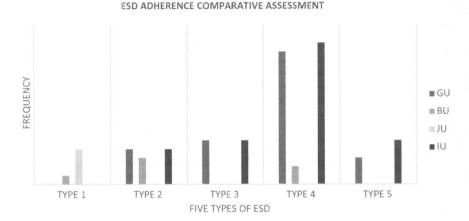

Figure 7.2 Results from the application of the 'Five Types of ESD' analytical model to case studies based on the presence or absence of the components in courses/curricula.

represent and compare the levels of ESD adherence according to the comparative assessment in Figure 7.2.

References

Bowen, GA 2009, 'Document analysis as a qualitative research method', *Qualitative Research Journal*, vol. 9, no. 2, pp. 27–40.

Bridge, G, Bouzarovski, S, Bradshaw, M, & Eyre, N 2013, 'Geographies of energy transition: space, place and the low-carbon economy', *Energy Policy*, vol. 53, pp. 331–340.

Haigh, M 2014, 'From internationalisation to education for global citizenship: a multi-layered history', *Higher Education Quarterly*, vol. 68, no. 1, pp. 6–27.

Hein, G 1991, *Constructivist learning theory*, viewed 20 June 2018, www.exploratorium.edu/ifi/resources/constructivistlearning.html.

Henderson, H & Ikeda, D 2004, *Planetary citizenship: your values, beliefs, and actions can shape a sustainable world*, Middleway Press, Santa Monica, CA.

Kolb, DA 1984, *Experiential learning: experience as the source of learning and development*, Prentice-Hall, Upper Saddle River, NJ.

Mezirow, J 2009, 'An overview of transformative learning', in K Illeris (ed.), *Contemporary theories of learning*, Routledge, Abingdon, pp. 90–105.

Scott, WAH & Gough, SR 2003, *Sustainable development and learning: framing the issues*, Routledge Falmer, London, pp. 113, 116.

St. Clair, R 2003, 'Words for the world: creating critical environmental literacy for adults', *New Directions for Adult and Continuing Education*, vol. 2003, no. 99, pp. 69–78.

van der Horst, D, Harrison, C, Staddon, S, & Wood, G 2016, 'Improving energy literacy through student-led fieldwork – at home', *Journal of Geography in Higher Education*, vol. 40, no. 1, pp. 67–76.

Vare, P 2007, 'From practice to theory: participation as learning in the context of sustainable development projects', in AD Reid, BB Jensen, J Nikel, & V Simovska, (eds.), *Participation and learning: perspectives on education and the environment, health and sustainability*, Springer Press, Dordrecht, pp. 128–143.

Vare, P & Scott, WAH 2007, 'Learning for a change: exploring the relationship between education and sustainable development', *Journal of Education for Sustainable Development*, vol. 1, no. 2, pp. 3, 4.

Walkington, H, Dyer, S, Solem, M, Haigh, M & Waddington, S 2018, 'A capabilities approach to higher education: geocapabilities and implications for geography curricula', *Journal of Geography in Higher Education*, vol. 42, no. 1, pp. 7–24.

8 Indigenous and non-indigenous students' perspectives

We don't leave anything to the nature, we don't destroy or show that we have been there, this is our mantra (Else)

8.1 Introduction

This chapter outlines the findings arising from the interviews conducted with the students from the four case studies. The students' perspectives are essential to provide contrasts with the data collected from the lecturers' interviews and the documentary findings presented in the previous chapters. In fact, students showed interesting but different perspectives about the concept of Sustainable Development (SD) and energy literacy (EL) and expressed varied viewpoints about the courses analysed with different implications on learning and pedagogies for the Arctic courses, as explained in Chapter 10.

Table 8.1 A summary of the students' code names, educational institutions, countries, and institution population type

Student code name	University	Country	Institution population type
Anita (non-indigenous)	Borealis University	Norway	Non-indigenous
Ingrid (non-indigenous)	Borealis University	Norway	Non-indigenous
Odd (non-indigenous)	Borealis University	Norway	Non-indigenous
Anette (non-indigenous)	Borealis University	Norway	Non-indigenous
Karoline (non-indigenous)	Borealis University	Norway	Non-indigenous
Astrid (non-indigenous)	Borealis University	Norway	non-indigenous
Else (indigenous)	Glacial University	Norway	Indigenous
Mona (indigenous)	Glacial University	Norway	Indigenous
Hilde (indigenous)	Glacial University	Norway	Indigenous
Helene (indigenous)	Glacial University	Norway	Indigenous
Reidun (indigenous)	Glacial University	Norway	Indigenous

(*Continued*)

Table 8.1 (Cont.)

Student code name	University	Country	Institution population type
Margit (indigenous)	Glacial University	Norway	Indigenous
Arthur (indigenous)	Juniper University	Canada	Mixed
Alice (indigenous)	Juniper University	Canada	Mixed
Florence (indigenous)	Juniper University	Canada	Mixed
Isaac (indigenous)	Juniper University	Canada	Mixed
Sam (non-indigenous)	Juniper University	Canada	Mixed
Emilie (non-indigenous)	Juniper University	Canada	Mixed
Rylee (non-indigenous)	Ice University	Alaska	Mixed
Evan (non-indigenous)	Ice University	Alaska	Mixed
Cora (non-indigenous)	Ice University	Alaska	Mixed
Micah (non-indigenous)	Ice University	Alaska	Mixed
Chloe (indigenous)	Ice University	Alaska	Mixed
Clara (indigenous)	Ice University	Alaska	Mixed

The students from the different universities were identified under code names presented in Table 8.1 and they belong to different and varied ethnic groups (indigenous and non-indigenous).

The data were synthetized in order to create a typology to visualize which aspects were in evidence from the sample. The contexts in the four case studies present more dissimilarities than common points, as remarked in Chapter 4, where the coding of the lecturers was presented. The codes that emerged from the data collected from the students' interviews were the following:

a) Intergenerational dimension
b) SD as practice
c) Critical thinking

It was clear that students (indigenous and non-indigenous) understand the main characteristics of SD and energy, but they demonstrate different levels of experience in relation to these notions. These different levels of experience have implications as to how meaningful these notions are to them and how meaningfully they can apply their knowledge. For some, these are more theoretical discussions, but for others, SD and energy have practical impacts on their everyday lifestyles and on their well-being, now and in the future.

8.2 Intergenerational dimension

The students from Glacial University (GU) understand Sustainable Development (SD) according to cultural and societal perspectives. They commonly associate the concept of SD as intergenerational well-being or as 'growth to build a better life'. In other words, 'development' is associated with well-being and a better life for the family, the community and future generations. It also means the opportunity to exercise their traditions and keep their way of life and culture alive in the long term. We can understand that development is 'the opportunity to be indigenous'. This reveals an important *functioning* of Nussbaum (2011, p. 33) as discussed in Chapter 3 (see Section 3.3). The communitarian aspect is present in the discourse, suggesting that this is a process that happens at a collective level echoed by Delors et al. (1996, p. 37) and Haigh (2014a, 2014b).

The indigenous students expressed that SD is central to their course and it is connected to the value of nature as a provider of resources (e.g. food and shelter) and the importance of conservation was emphasized. Some students mentioned several times the importance of protecting nature, as a place that people leave as clean as possible for future generations.

ELSE (INDIGENOUS): Nature is a place that you leave as clean as if you haven't ever been there. It is stupid to use all the nature, it is not our way of thinking. Nature is to be cared; we see more value on untouched nature.
HILDE (INDIGENOUS): Sustainable Development means how we can use it without destroying it and how we can keep it the same. I think I want to keep the nature for my kids, and I want my kids to keep our traditional way of life and I don't want them to stop it because of lack of land or because of climate change.

Ideas related to consumerism as opposite to development emerged as well as how to use nature without destroying it and leaving it to younger generations. The intergenerational dimension has been mentioned several times suggesting that it is a central aspect of development for the students at GU. The time dimension of SD is a cultural characteristic for the GU students:

REIDUN (INDIGENOUS): The difference is that other people think about the money and how much they can take from the nature and use it for themselves. We think about nature and how we can use it without destroying it and how our grandchildren will use it and how they can live in this place in the future.
MONA (INDIGENOUS): I am not positive because companies and people outside our culture try to build things, try to explore energy and minerals, they don't think about nature and the traditional way of life, our traditional activities, they only see money there. It is important to start thinking how we can make energy without destroying everything around us.

Some students also mentioned the dilemma between modern life and a traditional way of life and the importance of ESD as a way of influencing people to think how they can shape and impact positively the development process by considering more than one perspective in the educational discourse, more precisely these students referred to the inclusion of the indigenous perspective to be able to understand different sides of the discourse:

MONA (INDIGENOUS): Development means to make life easier and better; it is also to create opportunities to use the traditions to keep the old way of life and to protect the nature. ESD, may be teaching people to think how they can impact on development; may be changing the points of view about development because most people think that building big cities and shopping centres is development, but it is not. Explaining the consequences of choices, and making people think and act is the most important thing for me in education, start thinking like that.

The intergenerational dimension was also mentioned by the students interviewed at Borealis University (BU) but it was visible that the meaning of this component is a bit distinct. The concept of SD is perceived by the students at Borealis University (BU) as a development with progress but with no damage to younger (future) generations and it has been associated with clean energy resources.

ANITA (NON-INDIGENOUS): Sustainable Development means progress with no damage to younger generations; it means no use of oil but using clean energy.
INGRID (NON-INDIGENOUS): I feel that Sustainable Development means to produce energy that does not cause harm to the nature or to people after us.
ODD (NON-INDIGENOUS): Energy should not be got from a mean that destroys everything else; example would be oil drilling that would destroy our source of food for the future and for now as well. This is not sustainable.

It suggests a clear trend of associating the concept of SD to production and use of renewable energy sources, envisaging a better world for future generations by reducing the exploration and use of fossil fuels. This position is primarily based on the international documents' provisions[1] and on awareness originated by the experience from international funded projects conducted by some of the lecturers interviewed in this research. The application of knowledge is desired, but it is not measured, as the students return to their home countries but there is no certainty if they will apply the knowledge in the field.

INGRID (NON-INDIGENOUS): It means to use resources not causing consequences for the future generations.
ODD (NON-INDIGENOUS): It means to be able to provide for ourselves now, but not diminishing the possibility for the next generation to provide for them in the same way.

In the case of the Borealis University (BU), students' reports are oriented to technical views on SD and energy. The focus of the course is on effects of climate change and 'energy and environment' these are central components of the course according to the students' reports. SD is studied taking into consideration these biogeophysical main elements (impacts) and the interconnections among them.

Students like Emilie (non-indigenous) and Florence (indigenous) from Juniper University (JU) also emphasized the idea of durability (the time component or intergenerational dimension) in terms of SD practices, as follows:

EMILIE (NON-INDIGENOUS): My idea of Sustainable Development is some sort of development practice that can be long lasting and ensure that future generations can take advantage of what we are able to use currently; basically, something long lasting, not temporary.

FLORENCE (INDIGENOUS): The course helped me to understand even more how sustainability is built within the intrinsic nature of culture; about maintaining the relationship with the environment in a way it sustains the whole Earth *and the next seven generations, and the next generation has to grow up the next seven next generations; the future is always in focus.*

Florence demonstrated to have a real benefit from the course as an indigenous student because based on her innate and intuitive Traditional Knowledge she can critically visualize two different worlds and take her own conclusions about SD and the time dimension of it. She also called attention to the valuable resource of TK and the collaborative spirit of communities in sharing this for the benefit of future generations:

FLORENCE (INDIGENOUS): I think it is really important that people talk to indigenous peoples of the lands, because these people sustained these lands for thousands of years and to such a short of a time, we came to this point of being sick as it is, due to energy development most and the ways people use them. There is a lot of knowledge that the indigenous people have that is accessible if you just go to them because they want to share it. I know in Canada, indigenous people are the first in trying to protect the environment. They are saying 'we need to be able to sustain the Earth' and 'we need to care of the generations after us' and I think as a society of consumers we do not really think how we are affecting our grand kids, and that might be important to sit back and think about that and gain some information from the people that have been doing this for thousands of years.

The intergenerational dimension was also mentioned by the students from Ice University (IU) in the interviews:

CLARA (INDIGENOUS): Sustainable Development is to use the resources we have while making sure the future generations will be able to use those resources so I think development should respect the nature.

MICAH (NON-INDIGENOUS): Sustainable Development would be building with the future in mind – something that does not need outside energy to continue in perpetuity.

CORA (NON-INDIGENOUS): Sustainable Development means a development that does not destroy the environment and takes into consideration keeping the Earth clean and healthy to be passed to other generations.

8.3 Sustainable Development (SD) as practice

SD understood as practice has several additional implications. It involves autonomy, empowerment, freedom, and renewable energy resources under the viewpoint of students interviewed in the four universities.

At GU the practical component is part of the learning experience.

ELSE (INDIGENOUS): Everybody here experiences climate change every day, because this year we had a very warm and unusual autumn. We have had only minus 4 degrees in November and the ice and lakes were not frozen. It has a big impact in our local activities; we don't have industries here in the community, but we suffer with the world commercial madness.

Karoline from Borealis University (BU) reported that she would like the course to focus more on Sustainable Development as practice, as follows:

KAROLINE (NON-INDIGENOUS): The course should be more focused on practice and field work; this way we see the effects of climate change and energy production; it is just for us to learn more about the environment we live in and to see the landscape because it is easy to see it on the books we read; it is better if the teacher shows us in real life how it affects areas.

The students interviewed at Juniper University (JU) were mixed, meaning that non-indigenous and indigenous students participated in the interviews and they revealed different approaches to SD and energy. In general, indigenous and non-indigenous students reported that the course presents a significant amount of information related to SD; however, SD is not a central component of the course. They believe that SD should be more emphasized, but they also realized that the course has shifted the focus to environmental aspects (environmental impacts of operations) and human interactions with the environment.

SAM (NON-INDIGENOUS): Sustainable Development is not central. There are aspects when Sustainable Development is touched on, but I would say that it is not the focus. It is more about providing all the technical knowledge and getting it to the world and then have some Sustainable Development aspects on the side.

Sam and Emily (non-indigenous students) refer to SD as a development that satisfies the three pillars of the TBL approaching the subject under the viewpoint of theories, international treaties, technology, and environmental management. Sam mentioned an example of a local extractive company, its technologies and the percentage of annual profits invested on the local environment.

Students like Florence (indigenous First Nation) also emphasized the long-term view of SD when arguing that development means using resources in a balanced way so we can maintain the Earth for a long period of time. However, other students like Arthur and Isaac, both indigenous First Nation, showed a different position in relation to the concept of SD emphasizing the lack of practical application of the concept:

ARTHUR (INDIGENOUS): I think the term Sustainable Development is just a good term that is used; it does not really mean much to me. It does not really mean anything to me. It is a formal term and it seems not to mean much to people.

ISAAC (INDIGENOUS): Sustainable Development to me feels like, at this point, just words. I wish that we could develop sustainably but I don't see it happening right now specially in the future, I don't see how it comes to fruition.

Arthur and Alice (both indigenous students) also associate the notion of SD to using resources other than fossil fuels and moving into solar energy and other alternative energy resources. He also emphasized that he does not consider the style we currently live as sustainable:

ARTHUR (INDIGENOUS): The most important idea about Sustainable Development is to move away from fossil fuels, to move into solar panels and more diverse energies, but I don't see this style we live as sustainable, it is what it means to me, it is like theory, sci-fi. I don't see anything practical.

ALICE (INDIGENOUS): I think it would really benefit us to look at other avenues like starting using solar panels, renewables, and may be starting teaching people how to fix their own homes, so they have this knowledge and instead relying on the government we have, things like that. Learning for practice, preparing for the future and it is not just us; we need to look after we need to look after of the future people as well. There is a lot of adaptation to be made.

Practice also seems to be the approach of Alice (indigenous) in relation to SD concept when explaining that where she lives (in an indigenous reserve) there is a lack of materials to build and to fix the houses adequately to face the rigorous winter. She also mentioned that in a neighbouring community there has been investment in a new solar energy system (mini-grid with solar panels) and it is helping the community to save money on energy bills (expensive diesel-based generators) that can be used to supply other community needs. The notion of SD is perceived as practice, autonomy and renewable energy resources by the indigenous students according to their cultural and communitarian experience (from the indigenous reserves and communities) but this is the aspect that lacks in a more formal and theoretical presence in the course.

The students at Ice University (IU) also mentioned the UN definition of SD and the importance of conservation and the balanced use of resources allowing future generations to be able to use those resources as well. However, the crucial element is the practical aspects of everyday life. This was Clara's opinion endorsed by Micah and Cora who emphasized the intergenerational dimension of the concept adding that energy should not come from external sources like oil that needs to be extracted to be used, but from the nature like solar energy. Energy literacy for these students is a means of making the SD concept more practical, because renewable energy has a real and practical impact in making their development process and lives more sustainable, more autonomous and in creating well-being for their communities at present in the future.

RYLEE (NON-INDIGENOUS): When you think about Sustainable Development specially in the Arctic, you must think on how to live on the moon, how do you make it sustainable on the moon you need to think in the same way, how do you make it sustainable in the Arctic.

Rylee (non-indigenous) and Evan (non-indigenous) expressed their notion of SD related to very practical aspects of everyday life. They both emphasized that the development they experience locally is not sustainable. This is because they depend on resources that come from outside. They do not have local autonomy.

RYLEE (NON-INDIGENOUS): (…) when I think of Sustainable Development, especially in the Arctic, it blows my mind because you have to think in terms of sustainable food. Our food is trucked from outside and it is flown or barged from outside, everything comes to us from outside and it gets here from who knows where and when there is a flood or a transport issue we do not have food in the grocery store (…) we are in the end of the line and it is really noticeable how reliant we are on this supply chain that is so tenuous and not a sustainable option. *I live in Alaska where all development is completely not sustainable.*

RYLEE (NON-INDIGENOUS): We have farms in the summer and farmers can store their food and they are self-sufficient, the rest of us feel incapable of growing, but we fish. I fish in the summer and I have fish all winter, we hunt and usually we have some kind of meat but bread, flour any of that comes from elsewhere (...) the farming we have here is tiny comparing to the number of people we need to feed (...) when you think about sustainable development specially in the Arctic, *you must think on how to live on the moon, how do you make it sustainable on the moon; you need to think in the same way, how do you make it sustainable in the Arctic.*

In the same line, Evan pointed out the importance of the economic component of the SD concept, as it is difficult to have SD when people are dependent on external goods and services and they do not have the capabilities to be autonomous or change their reality. Development that is sustainable means to have a functional local economic activity and access to resources (energy inclusive) to resolve practical daily issues and needs:

EVAN (NON-INDIGENOUS): *Sustainable Development is not possible where you live dependent of goods and services brought from the outside*, if you are talking about developing an area, it needs to be a place where people have some sort of finished goods and process, procedures, technical skills they can count on, and they can depend on to go forward from generation to generation. Most communities do not have an economy of any sort, and technology, building materials and transportation require money. If we agree to have a functional local economy, we will need to train people so they can actually produce and use materials themselves. So, for me sustainable development means to advance in new technologies, techniques and materials and provide mechanisms for people to know how to have access to resources and to know how to use resources.

The students from Ice University (IU) made reference to the fact that SD is central to the course they study, and sustainable energy is ingrained into their SD approach and understanding of the SD concept, that implies a practical approach to their learning experience through real-life energy projects, e.g. smart grids. SD and sustainable energy are important for these students due to the practical applications in everyday life issues.

8.4 Critical thinking

Critical thinking was a very relevant aspect mentioned by several students from the four case studies. At GU, Mona (indigenous) was emphatic about the importance of 'preparing people to think' and 'to understand contexts of reality'. She reported that the course had a huge impact on her because of the following reasons:

MONA (INDIGENOUS): I am proud of myself because I built up my identity along the years through the indigenous tradition and through the studies here because I started to understand other perspectives and what is really important to my grandparents, to the generation before them and it is not minerals or energy but land, water and freedom.

Now we are so dependent of energy, computers, internet and we are so fast; we cannot come out of this dependency. The course taught me how to think and how I will teach people to think; to ask questions and to make the studying process more practical not telling facts but making people to research, to read, to make their own reflections based on a different perspective.

Else also emphasized the importance of the methodology to her critical thinking process as she looked at issues under two different perspectives (indigenous and non-indigenous). This is an aspect that Mona expanded when explaining the importance of looking at issues from the viewpoint of TK:

ELSE (INDIGENOUS): I can see differences in the methodology here compared to other universities because we are looking both in an indigenous way and looking at the western world. We see the values, what others wrote, and we don't have right or wrong. One capability I acquired is to reflect about my own world.

MONA (INDIGENOUS): I research with people, there are talks, audio files, pictures, books and the learning from generations to generations. I need to read, I need to enquire [from] people. You try and fail, you make your own reflections in what you did wrong and try again. You give students the opportunity to try and make mistakes and try again, then they start to think.

Anette (non-indigenous) from Borealis University (BU) mentioned the importance of critical thinking in her interview when emphasizing the importance to understand the Earth's systems, a point that was also mentioned by Karoline (non-indigenous) and Odd (non-indigenous) suggesting a 'systemic thinking' approach:

ANETTE (NON-INDIGENOUS): There are so many factors affecting the Earth and in connection. Most part of them are because humans affect the climate and the climate and landscape affect us.

KAROLINE (NON-INDIGENOUS): It is important to understand the process behind each phenomenon – the processes and the links to other processes, to see the big picture and learn different methods to get results.

ODD (NON-INDIGENOUS): In general, at the university, I learned to be critical to every source I read; how to write and read texts and I learned how societies function; and how they function in relation to climate change.

Florence (indigenous), a First Nation student from Juniper University (JU) made interesting comments in her interview. Florence's report brings evidence of the critical thinking and personal empowerment generated by the encounter of different types of knowledge; academic knowledge and TK, as follows:

FLORENCE (INDIGENOUS): I think there is a combination. There are lots of things I learned growing up that I didn't realise until I was sitting the course. I have learned things about sustainable development in the course and I have learned things that are unconscious and became conscious with the course. I have experienced the correlations between these different types of knowledge.

Emilie (non-indigenous) emphasised critical thinking as the most important skill she developed by studying the course connecting it to the importance of reading from reliable sources and connecting different aspects of the topic. She uses critical thinking to question the true meaning of sustainable development.

EMILIE (NON-INDIGENOUS): My critical thinking skill is the biggest skill I developed in human geography – to be critical and understand the components. The course is also about research and being critical on what people read; it is also looking at all the different facets.
EMILIE (NON-INDIGENOUS): I think the whole discussion on sustainability is really interesting and I feel that my knowledge that emerged in the last 3 years there were some projects that focused on it but at the same time I feel that it is such a big overarching topic and what does it really mean?

At Ice University (IU) Evan (non-indigenous) and Cora (non-indigenous) made important points about the importance of critical thinking and the power of testing concepts, approaches, methods, and practices when saying:

EVAN (NON-INDIGENOUS): You are in a place where people are operating in the limit, this is the best place to reflect and test concepts and terminologies.
CORA (NON-INDIGENOUS): Knowing more about energy, I can make an intelligent decision regarding energy efficiency and renewable energy choices now.

8.5 The students' perception of the key objective of the courses as learners at the four case study universities

8.5.1 Glacial University (GU)

The students' perception of the key objectives of the courses as learners showed relevant evidence confirming points presented in the coding sections above.

Students from GU reported their perception regarding the objectives of the course as focused on documenting TK through research, developing critical thinking and reflection on the details of every issue.

MONA (INDIGENOUS): Making me, forcing me to research information, to think by myself, to make conclusions, to reflect on every single detail, every single word.

MONA (INDIGENOUS): (…) you have to research, you read a lot of books and you have to be critical and if you say something in class the others will be critical and ask, 'Are you sure?' I am sure because I can say this way because I have this kind of research, it really makes us think (…).

ELSE (INDIGENOUS): I learnt from this course things about Linda Highsmith that is a New Zealand very important indigenous researcher. She has worked with taking our way of life back again after the assimilation of indigenous people.

8.5.2 Borealis University (BU)

Annette identified as a course objective the skill to *understand systems* on Earth and interconnected factors caused by humans affecting climate and landscape:

ANETTE (NON-INDIGENOUS): Understand systems on Earth.

ANETTE (NON-INDIGENOUS): Yes, but it should be clearer but there are so many factors affecting and in connection, and most part of them are because humans affect the climate and the climate and landscape affect us.

Odd mentioned about the importance of connecting his learning on geophysical to social systems and the course objectives tending to focus on physical geography and climate, as follows:

ODD (NON-INDIGENOUS): It is the study of physical geography the study of landscapes and climate, but these things go together, and I undertook modules on social studies in geography talking about renewable cities, so the course was focused on physical geography and climate.

Karoline (non-indigenous) emphasized, as an objective of the course, the understanding behind the phenomenon:

KAROLINE (NON-INDIGENOUS): Understanding the process behind each phenomenon and why things look like how they look like.

8.5.3 Juniper University (JU)

Alice (indigenous) and Isaac (indigenous) pointed out that the course is more focused on *environmental management or management of natural resources* while Arthur reported that the objective of the course is to increase awareness of alternative sources of energy and provide a technical knowledge base, but it will not be fully realized before the indigenous perspective is completely understood. This emphasizes the importance of learning under more than one perspective (indigenous and non-indigenous):

ARTHUR (INDIGENOUS): The key objective is not only to increase awareness of alternative forms of energy but to ensure that these technologies get researched and continue developing.

ARTHUR (INDIGENOUS): Yes, it is linked to technology, but I don't think it will happen until indigenous perspective is completely understood.

ALICE (INDIGENOUS): The course is focused on learning how to manage natural resources.

ISAAC (INDIGENOUS): The objective of the course is to look at how humans interact with the environment, exploring aspects from all over the world. Another objective is the management of natural resources; it is more practical, and we look at law, policy, historical perspective, and contemporary perspectives.

Emilie (non-indigenous) emphasized the connections among factors and the importance of linking the knowledge with the fieldwork experiences. Emilie understands and expresses the importance of practice:

EMILIE (NON-INDIGENOUS): I think the objective is to provide an understanding of real-life situations. You learn about zoning and concepts that give you the background to real-life experiences. This is what I like about it, the knowledge about the interaction among people and their built surrounding environment and the connections among factors. You don't necessarily think that there is a connection but actually when you learn you consider the connections and the relationships.

Florence (indigenous), on the other hand, mentioned that there is an inner contradiction in the course because according to her view there are two sides, the 'sustainability people' versus the 'energy people':

FLORENCE (INDIGENOUS): There are two sides. On one hand there is this very environmentally conscious community that teaches you a lot of content on conservation and sustainability but, on the other hand, there is funding from the energy industry. There is an inner contradiction.

8.5.4 Ice University (IU)

Clara (indigenous) mentioned that the course helped her understand energy production and use, the concept of green building, and quantifiable ways to determine what is sustainable and concrete elements to make a viable decision about SD.

CLARA (INDIGENOUS): I became more fluent and my Sustainable Energy Literacy is improved; I can talk about KWh, I know the difference between power and energy, I understand energy production and use, the concept of green building, I know concrete things to make a viable decision about SD, quantifiable ways to determine what is sustainable, how to switch things to make it more sustainable and these are things that I learnt in class.

Cora (non-indigenous) mentioned as one objective of the course to provide knowledge about energy efficiency connected to new technologies and Evan emphasized that the course approaches adaptability and the perspective of researching ways of making use of renewable resources and applying it in real world settings at affordable prices to locals. An additional aspect is the technical skills to allow students to make a proper use of these technologies and be able to resolve real world problems.

CORA (NON-INDIGENOUS): I think the focus is on energy efficiency and the new technologies, renewable energy new technologies.

EVAN (NON-INDIGENOUS): Approaching sustainable energy from the perspective of researching ways of making use of those resources and from the perspective to find ways to apply it in real world settings at affordable price to locals and finding ways to apply technical skill training to the locals, so they can make proper use of it.

8.6 Capabilities developed from a student perspective

8.6.1 Glacial University (GU)

Students from the GU reported that they acquired skills related to reflection, critical thinking, and research, but mainly multicultural worldview. This evidence raises the point of 'knowing how to live together' (Delors et al. 1996, p. 37), considering different perspectives (indigenous and non-indigenous) when learning as per the important attributes of being an Arctic citizen.

ELSE (INDIGENOUS): I learnt how to reflect about my own world (...) it has built my world view and I also can tell about cultures, how people have been living so far.

8.6.2 Borealis University (BU)

The capabilities that students feel they developed as a result of taking this course at Borealis University (BU) are related to the capacity to reflect, to develop critical thinking, analytical skills and act with responsibility, as well as identifying the correlations between physical, climatic and societal elements.

Anita (non-indigenous), for example, identified critical thinking as a main capability acquired. She mentioned she exercises a more critical and responsible reflection to inform her decisions in relation to these geographical contexts:

ANITA (NON-INDIGENOUS): I think I reflect more, about my life and peoples' actions. I feel a lot of responsibility and I feel angry when people don't care. Critical reflection and responsibility are key elements.

For Ingrid (non-indigenous), Odd (non-indigenous), Anette (non-indigenous), and Karoline (non-indigenous), critical thinking is also associated with evaluation of research sources and methods, how to interpret the literature they read (mainly the causes and consequences of phenomena) and how to connect ideas and different processes.

INGRID (NON-INDIGENOUS): I don't feel that I am so different after that, I feel that I have more knowledge about what is causing effects and kinds of energy sources.
ODD (NON-INDIGENOUS): I learnt to be critical to every source I approach. I learnt how to write and read texts in geography. It gave me a clear understanding of how landscape works and how the physical process works and how climate change and landscapes relate and how societies function and how they function in relation to climate change.
ANETTE (NON-INDIGENOUS): Skills to understand the causes of phenomena and analytical knowledge and skills.
KAROLINE (NON-INDIGENOUS): To see the processes and the links to other processes, to see a big picture and learning different methods that we can use to get the results afterwards and learning GIS, for example, is an important thing to learn these days.

8.6.3 Juniper University (JU)

Students from Juniper University (JU) demonstrated to have different views about the capabilities acquired through the course. Indigenous and non-indigenous students were interviewed. Arthur, an indigenous student, identified important attributes like leadership, confidence, and critical thinking as relevant characteristics a student may have in order to apply knowledge. However, when asked about the capabilities Arthur acquired as a learner by doing the course, he mentioned positive points but also the negative side of being in the university:

ARTHUR (INDIGENOUS): I don't believe that the course is really tailored to create attributes in people, I really think that you take the course just to get the grade and that's it. I really believe that you get these attributes from your community and from your own life experience. I don't believe that communities come here to get these skills. There are positives to be in the university. You learn things that you would not learn out there, but it is just hard sometimes because there are different environments and we don't feel that we can share our own opinion, or our learnings. Here we have to compromise, to negotiate, to live this educated life in order to be someone, so this is a big misunderstanding, a big mismatch, so we are almost here because we have to be.

Critical thinking and the ability to explore the interconnections and facets of knowledge were skills that students like Emilie also emphasized:

EMILIE (NON-INDIGENOUS): My critical thinking skills are probably the biggest skill that I have developed in human geography, the ability to question things and not to accept things of face-value, because things are not so obvious and there are so many factors involved; but to be critical and understand the components. You think that critical thinking is a skill that everybody has, but it is not. Human geography is about research, it is very important in our department and it is about finding the relationships and to be critical to others' research too. It is about to be critical on what you read and to look at all the different facets.

Florence (indigenous) called attention to something that she called 'geographical consciousness' as a consumer as an important capability. She emphasized the big footprint consumers have in urban areas and the fact that people do not think about those communities living in sensitive and remote areas like the Arctic. She emphasized the point that no one thinks how actions, decision-making, and the development process in other areas of the world affect the Arctic. This revealed the important evidence of her global perspective, her sense of Arctic and global citizenship as follows:

FLORENCE (INDIGENOUS): I have a great consciousness of what a big footprint I have and also like when I am choosing to buy certain items, when I am choosing as a consumer to go into the road and purchasing how it affects different people. People in urban centres do not really think about how their actions affect those people that live in areas like the Arctic, and when it comes to my family in the Arctic, people don't think about them, it is so far away, people don't think about that all the time, they don't really understand how aggressive climate change is and how development has affected those areas and so when it is out of your sight you don't really think about it, so I think just realizing how important sustainable development is I have gained the consciousness about my decisions and how they affect other people, even if it is not just here around me or even the Arctic, in choosing something I buy.

Students mentioned that they acquired skills related to management of resources, practical skills to be used in jobs like writing reports, preparing environmental assessments under political, legal, sociological, and economic aspects giving evidence that employability is an important course guideline.

8.6.4 Ice University (IU)

The students at Ice University (IU) have a specific profile. They are of mixed ethnic groups (indigenous and non-indigenous), they vary in terms of age, as there are young students but also a great number of mature students that live across the Arctic region including remote areas; they have different professional backgrounds and the level of applicability of knowledge and skills are higher because students attend university courses with the intention to learn and develop skills to be applied immediately in the field. The environment (climatic and socio- economic environment) is harsh in Alaska and it demands the immediate application of the skills acquired. In general, 'the field' is their homes and professional projects developed in urban and remote areas.

Clara, for example, took the course and her intention when studying was to apply knowledge and skills to resolve practical problems of her everyday life. She demonstrated excitement when saying:

CLARA (INDIGENOUS): I am so excited about that because Cora is my neighbour and she has the same issues about energy, ventilation and mould that I do and I am so glad to have learnt these things for my assignment project and glad to be able to test my knowledge into practice to solve these things in my own home and share information as much as possible with other people and galvanize builders, developers, people that do home assessments, and figure out how we can help and correct all those issues in practice.

Practice and problem-solving seem to be the words when referring to Ice University (IU) learning experience. Rylee (non-indigenous) emphasized that the course provided him the capacity to look at problems and do a better evaluation on a technical basis. He also emphasized the relevance of energy efficiency as a very important component of sustainable energy because it is not only a matter of generating more energy but saving what you have as a crucial point at dealing with energy problems.

RYLEE (NON-INDIGENOUS): As an engineer, it gave me the capacity to look at a problem and evaluate that problem in a lot of different ways and really understand what is going on more than someone else that does not have a technical background and cannot go into the details of what is going on in an energy problem. My focus is efficiency; I don't look at alternative energies, I look at what I have, and I make it better.

Micah (non-indigenous) also mentioned the importance of interacting number-wise by translating ideas into figures by quantifying and measuring energy consumption of different devices as well as the financial and carbon costs involved in energy problems. It seems an important capability in energy literacy in order to have ideas translated into figures and knowing how to structure a project in terms of energy capacity, investment and, for instance, the number of solar panels that are required to realize the project.

MICAH (NON-INDIGENOUS): Being able to think through the maths of things by putting ideas into numbers and see how they are going to interact number wise; How much energy I am using with different devices at the time, taking the KW and calculating the device uses and the amount of watts and how much this amount of energy will cost me and how much oil it will consume. The course gave me a perspective about my footprint and how much oil I am using and how much carbon ends up in the atmosphere because of that, the numbers into the system is the add up of this course.

EVAN (NON-INDIGENOUS): I think this is exactly the perspective of the course, because, basically, you are in a place (the Arctic) where people are operating in the limit. This is the best place to test these concepts, these terminologies.

Evan (non-indigenous) mentioned an important capability that is 'the capacity for sustainability'

EVAN (NON-INDIGENOUS): the capabilities developed, I would say that it would be awareness, and capacity for sustainability.

This capacity for sustainability according to Evan (non-indigenous), relates to managing resources on a macro scale by assessing power consumption, by knowing how to provide insulation to buildings in order to save energy and use energy efficient devices and equipment to reduce energy consumption, costs and footprint. Evan also mentioned that it would be really important to apply this capability at an industrial level because inefficiency in industrial equipment is a concern in Alaska.

Cora (non-indigenous) highlighted another relevant capability associated to make intelligent decisions regarding energy.

CORA (NON-INDIGENOUS): because I didn't know about this before and I had no experience with renewable energy resources previously to the course. I feel I know more about the topic and I can make an intelligent decision regarding energy efficiency and renewable energy choices now.

8.7 The balance of the course from a student perspective – academic vs. vocational

The students of the four universities had to draw a line on a rectangular diagram to try to measure the balance of the respective course positioning the perpendicular line on the left, on the right or in the middle to infer the predominant feature of the course.

Figure 8.1 demonstrates the unanimous opinion about the academic characteristic of the courses analysed at the four case study universities. The variations in the opinions of the 24 students interviewed and shown on the diagram are linked to the students' backgrounds and expectations of a more practical approach to academic knowledge.

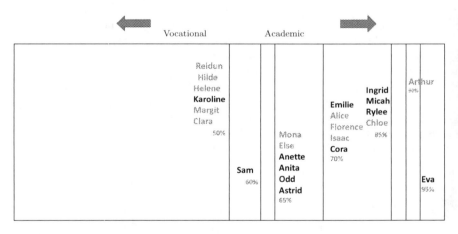

Figure 8.1 The academic characteristic of the courses according to interviewed students scaled by percentage, where 50% means 50% of the course is considered to be vocational and 50% academic; and 95% means 95% of the course is academic with only 5% vocational. The names in black represent the non-indigenous students while the names in grey represent the indigenous students.

In general terms, all students agree the courses are more academic, with those at Ice University (IU) being the most vocational. When interviewing the students at Ice University (IU) it was clear that the course is considered academic by the students, but at different degrees and with a relevant vocational component related to solar and wind energy projects in remote communities and in urban areas that require technical training how to operate these systems. Students mentioned the importance of the practical side of the learning mainly in remote communities because the locals could have better control of technologies and could be more independent of external technicians to do the maintenance and repair of solar panels and wind turbines.

Another example of the vocational aspect is perceived by the students from GU. They expressed very positive opinions regarding the balance of the course reporting that the learning acquired from the course is to be used in life not only in the workplace. They consider the course has a very good balance because it takes the indigenous life very closely to the education and the focus is on *life-long learning* instead of only knowledge for the workplace. Mona and Else illustrates this point:

MONA (INDIGENOUS): I guess it is more academic, this is learning to use in life not just for work. I have knowledge of how I should work but it is more than just preparing for workplace. The course is more academic, and the balance is really good I guess because you are prepared for the work, but you have more knowledge and you get the insight how you can develop yourself further. I would say it is 65% academic.

ELSE (INDIGENOUS): I think it is more academic and they are taking the indigenous life very closely to the education, so it is hard to say and draw a line there. I would draw a line in the middle.

Students from Borealis University (BU) and from Juniper University (JU) also mentioned they would have liked a more practical course. They mentioned that they would include more practice through field trips, additional experience, and training to enhance work skills. This perspective reflects the paradox academic versus vocational that several students in more westernized style courses experience in the HE, because students believe they need practical experience to apply for well-paid jobs despite the academic knowledge.

8.8 The process through which these curricula are designed and reviewed from a student perspective

8.8.1 *Glacial University (GU)*

The students interviewed at GU were asked about the course design and evaluation process, methods of evaluation, if their demands were attended to and if changes were promoted by the university as a result of this process. The reports confirmed that the curriculum is designed with the joint participation of staff and students based on the aspects evaluated in an end of term meeting.

MONA (INDIGENOUS): We are obliged to evaluate each course we take and then there are students' representatives here at the school board, so we can put our cases up there.

As a result of this method of evaluation students can express difficulties they had along the course. Mona illustrated this point by providing an example in relation to a specific online module delivered by an overseas lecturer and the communication difficulties the group had at that time. The students' representatives brought the case to the board and the lecturer was scheduled to deliver the course in person to the students. In a multicultural group, communication strategies are really important because of the multicultural and multilingual feature of the groups across the Arctic.

MONA (INDIGENOUS): We had an internet module with a foreign teacher, and we had difficulties to communicate with her, so we took this to the board and we finally had her in person to teach us here.

Another relevant result promoted by the evaluation process conducted at GU was related to pedagogies. Mona (indigenous) mentioned that the learning process at the university has always been oriented to 'learning by doing' (practice) and making inquiries (inquisitive) and making questions to understand the best ways of dealing with Traditional Knowledge (TK).

MONA (INDIGENOUS): It is more like learning by doing and asking questions. If I ask what the traditional way is of making fish skin, you need to research that, you go and read those books and we can talk about.

The latest evaluation process emphasised even more the relevance of research in the academic practice of indigenous students. This result was not exactly a change caused by students' complaints but promoted by a joint effort from lecturers and students to apply the most engaging pedagogies that demonstrated to have produced the best results for both. This has a direct application on how to design or re-design courses by taking the evaluation of previous terms as a means to improve the curriculum by co-creating the curriculum informing content, communication methods, and pedagogies appropriate to the indigenous style of learning.

8.8.2 Borealis University (BU)

The course evaluation system at BU consists of using a specific website where most part of the activities take place. Specific course readings and in-class tasks can be found in the website through which the course is evaluated. Anita mentioned the name of the website evaluation system that is kept confidential for ethical reasons. Anita also mentioned that the course evaluation can be made through a reference group, but the process of change is slow.

ANITA (NON-INDIGENOUS): We had this website where we are asked about the reading and in-class tasks. We have different courses and we evaluate them through the website, but the teachers don't care much about what we say.

ANITA (NON-INDIGENOUS): We had this reference group and if the student was asked if he/she was satisfied I would talk to the teacher, so in my group, I was the reference to the teacher and the students were not that much happy with what was done; there were a lot of critics. I felt like sometimes he was not listening to us.

Ingrid (non-indigenous) was another student who mentioned the reference student system of course evaluation. She also added that students had an innovative idea to engage more students to provide course feedback, however changes were not observed yet.

INGRID (NON-INDIGENOUS): You can be a reference student if you want to be; if you want to make a difference you can do that but there were some students, for example that created an App to evaluate field trips and I think it was very interesting and it is a good example how to be a good influence and make things better. I haven't seen any change yet.

ODD (NON-INDIGENOUS): In general, there is a questionnaire in the end of the term and usually there are the reference students, I have been one once, you meet the reference student about your concerns and she or he will take it to the subject (module) lecturer so it is anonymous and it works well depending on the lecturer of course.

Annette (non-indigenous) emphasized that students answer the feedback questionnaires in the end of the course, but no feedback is given during the course and Karoline (non-indigenous) also made the point that it would be interesting to have a mid-term feedback to raise issues during the course in order to produce change when the students are still doing the course not after they left the course.

ANETTE (NON-INDIGENOUS): We do questionnaires in the end of the course but not during the course.
KAROLINE (NON-INDIGENOUS): We don't have anything to say until the end and if we don't take the representatives, (we are two students as representatives for each course) to talk to the lecturers, so in general we don't do, we don't talk to them.
KAROLINE (NON-INDIGENOUS): No, we don't have a voice; we have a voice later but not now, during the course. So, the things we bring up really don't

affect us as we only bring them in the end of the course, but I feel I have a voice if I go and talk to one of the councillors, advisers. If we go, we have a voice in the geography department, we can be heard and have a voice but apart from that I don't think we do that much.

8.8.3 Juniper University (JU)

Most of the students from Juniper University (JU) were asked about if they had the chance to evaluate the course and if they felt they had a voice to promote real changes in the curricula. The answers showed that most of the students do a final term evaluation that is related to the individual lecturers' teaching ability and methods, but not exactly related to the curriculum and *change, in this system, is perceived as a very slow process due to academic bureaucracy*. According to Alice, a lot of adaptation needs to be made. Arthur (indigenous), Alice (indigenous), Florence (indigenous), Isaac (indigenous), Sam (non-indigenous), and Emilie (non-indigenous) expressed strong opinions about their experience and *curriculum review*.

ARTHUR (INDIGENOUS): Absolutely, I don't think so at all, *specially being an indigenous you don't have a voice*. The reason because it is this way it is because they don't understand our legacy and because people don't understand the conception and the whole process of colonization.

ARTHUR (INDIGENOUS): I don't really know if there is cultural shock but the whole system, the pressing system like the official conciliation committees represent a longer legacy, of the development of Canada throughout until 1996, they created schools where they put children where they took the indigenous from the reserves and they put them into schools and to teach them their ways just to teach them and force them to assimilate. I am talking about assimilation.

ARTHUR (INDIGENOUS): I think that if the whole intention of this system is to give indigenous voice and actually to give legitimacy to indigenous voice I think that it has to come from indigenous people and *I really appreciate that you are taking the time to hear our voices but I think all of this needs to come into action* definitely and I really believe that around the world indigenous groups started to talk.

According to the local educational system, the lecturer has the power to determine the curriculum. The course depends on the lecturer's content design, approaches, pedagogies, and academic ideologies. The lecturer has autonomy to design the course/programme and the students generally evaluate the lecturer. This system involving end-of-term on-line surveys for course evaluation, according to the students, does not promote considerable change.

FLORENCE (INDIGENOUS): We have these mechanisms, in the end of the course, these on-line surveys that come and you can evaluate the course content and the professors and all that stuff at certain scales. I am not sure how effective they are in terms of changing curriculum; I know that they are required to be done but not sure how it was used to change course, content or not, I am not sure what they do with that information after all. I wish I could give you a good example. I haven't seen any example of change.

ALICE (INDIGENOUS): Most of the questionnaires are oriented to the professors' abilities and I don't know about much change they actually have. I guess that more isolated communities they will benefit the most from sustainable resources because it is difficult to get, electricity, signal in those places where they need help more than in urban communities where I come from, but in the future maybe it is good to start learning how to use sustainable resources in these remote areas.

ISAAC (INDIGENOUS): Every module has an online survey about the professor that teaches the class. I feel that the change is very slow mainly because of bureaucracy in the academy and I have witnessed this several times. Here, the curriculum is a five years cycle and it takes long to change anything. To review the courses are a slow process.

SAM (NON-INDIGENOUS): We have course evaluation at the end of every course but that is more an evaluation of the professors themselves not necessarily the course. Students do not have a voice. We have the students' union, but the university does what it wants and what is best for the university.

EMILIE (NON-INDIGENOUS): Definitely, I did some evaluations but I don't feel that they do much, because here the courses are more dependent on the professor than on the course topic in itself so take the same class with different professors will be a completely different experience and in terms of changes I have been part of a list of graduate students and I have been a student representative for 3 years but I felt that I was unable to do anything so this university in particular human geography is under natural sciences department and our representations are very tiny and sometimes are under looked and it created issues, a lot of issues with a lot of students in relation to scholarships.

8.8.4 Ice University (IU)

The same question regarding course evaluation and curriculum review was asked to the students from Ice University (IU) in order to verify their perception on how the curriculum is reviewed and changed. Most of the students from Ice University (IU) emphasized that there is a survey at the end of the course and some of them expressed different opinions despite the final perception of the process to sound positive, there is awareness about the process of continuous development and improvement to enhance the indigenous students experience according to their own ways of learning.

CLARA (INDIGENOUS): I would like to say that lecturers do evaluation surveys in the end of the course. I wish the course was not so academic and inflexible because we need more practical knowledge to apply in the everyday life.

EVAN (NON-INDIGENOUS): I don't know if I have seen a change in a mind-set of a professor. I haven't seen student in a position to actively change how a department approaches a specific topic, or to influence a professor to change its course. Almost every course, undergraduate and postgraduate, in the end of the course, there is a questionnaire and, in general, we can say what were the strengths and weaknesses in class.

CORA (NON-INDIGENOUS): I have given an evaluation for the course because we do it every time before the end of the term and the course worked for me. I think that for some people it wouldn't but for me it really worked, and I had no problem with that. I think it is getting better, and over the years it got better, with the blackboard, the contents and the projects.

This chapter has outlined the students' perspectives in relation to the courses they studied in the four case study universities. The students' opinions were essential to provide a counterpoint in relation to the data collected from the staff interviews. Similarities in relation to the students' reports were identified and codified according to three specific patterns related to intergenerational dimension, SD as practice and critical thinking. All the students from these different locations called attention to the importance of practicing SD, in other words, they value the actions and behaviours with a practical impact on the well-being of the environment and communities not only in the short term but in relation to future generations. Students reported that the courses helped them to acquire the necessary theoretical knowledge and to develop critical thinking to assess and understand the (un)sustainability of their contexts. However, there are remarkable differences related to different cultures of learning, cultural expectations and different learning experiences across cultural groups that has a direct impact on levels of practice and knowledge applicability, curriculum design and review. The employability dilemma was also verified in Western-style courses (students' concern about job after the university) but it is not a priority in the indigenous learning experience more oriented to the cultural groups' well-being.

Note

1 (IPCC Special Report on Renewable Energy Sources and Climate Change Mitigation), Sathaye J, Lucon, O, Rahman, A, Christensen, J, Denton, F, Fujino, J, Heath, G, Kadner, S, Mirza, M, Rudnick, H, Schlaepfer, A, Shmakin, A, 2011, 'Renewable energy in the context of sustainable energy', in O Edenhofer, R Pichs-Madruga, Y Sokona, K Seyboth, P Matschoss, S Kadner, T Zwickel, P Eickemeier, G Hansen, S Schlömer & C von Stechow (eds.), *IPCC special report on renewable energy sources and climate change mitigation*, Cambridge University Press, Cambridge, UK and New York, NY.

References

Delors, J, et al. 1996, 'Learning: the treasure within', Report to UNESCO of the international commission on education for the twenty-first century, UNESCO, Paris, p. 37.

Haigh, M 2014a, 'From internationalisation to education for global citizenship: a multi-layered history', *Higher Education Quarterly*, vol. 68, no. 1, pp. 6–27.

Haigh, M 2014b, 'Gaia: "thinking like a planet" as transformative learning', *Journal of Geography in Higher Education*, vol. 38, no. 1, pp. 49–68.

IPCC 2007, 'Climate Change 2007: impacts, adaptation and vulnerability', in ML Parry, OF Canziani, JP Palutikof, PJ van der Linden & CE Hanson (eds.), *Contribution of Working Group II to the Fourth Assessment Report of the Intergovernmental Panel on Climate Change, 2007*, Cambridge University Press, 979 pp.

Nussbaum, M 2011, *Creating capabilities: the human development approach*. Harvard University Press, Cambridge, MA, pp. 33–34.

Sathaye, J, Lucon, O, Rahman, A, Christensen, J, Denton, F, Fujino, J, Heath, G, Kadner, S, Mirza, M, Rudnick, H, Schlaepfer, A & Shmakin, A 2011, 'Renewable energy in the context of sustainable energy', in O Edenhofer, R Pichs-Madruga, Y Sokona, K Seyboth, P Matschoss, S Kadner, T Zwickel, P Eickemeier, G Hansen, S Schlömer & C von Stechow (eds.), *IPCC special report on renewable energy sources and climate change mitigation*, Cambridge University Press, Cambridge, UK and New York, NY. 3–230.

UNCED 1992, *Agenda 21. UN Conference on Environment and Development (UNCED)*, UN Department of Economic and Social Affairs, New York, NY.

UNEP 2010, *Global trends in sustainable energy investment 2010. Analysis of trends and issues in the financing of renewable energy and energy efficiency*, United Nations Environment Programme (UNEP). Paris, France.

WCED 1987, *Our common future. World Commission on Environment and Development (WCED)*, Oxford University Press, Oxford, UK and New York, NY.

9 The relevance of the UN SDGs to the Arctic HE and curricula

9.1 Introduction

Adaptation seems to be the key word for the future of the Arctic. It is a powerful word based on a challenging and impermanent reality of continuous complex interactions between the natural and social structures that have been accelerated by the large-scale energy production and use in the last 200 years. These interactions are better understood when studied under a geographical (bio-regional) perspective, as some of these interactions, in extreme environments like the Arctic, can present dramatic consequences to indigenous peoples, but can also create extraordinary geo-capabilities for multicultural learners in multicultural spaces. In previous chapters, the Sustainable Development Goal (SDG) 7 (energy) was approached and it demonstrated to be a complex task to assess levels of energy literacy and adherence to Education for Sustainable Development (ESD). There are other 10 SDGs to be addressed in the Arctic.

9.2 The relevance of the UN SDGs to the Arctic

The geophysical and human aspects of the Arctic are naturally interlinked, and with globalization, the human aspects of our existence are even more integrated resulting in a range of collective opportunities, risks, and responsibilities as a civilization. Arctic communities have demonstrated specific adaptive strategies based on the encounter of Traditional and scientific knowledge and Traditional Knowledge documentation to deal with these complex contemporary challenges. This comparative study aimed at examining whether the multidisciplinary and multicultural perspectives of a sustainable energy model are being addressed by Arctic Higher Education (HE) curricula in times of carbon-constrained activities. The original contributions to knowledge of this study are the Arctic Pedagogical Model of Energy Literacy, ESD Adherence Parameters, and the Education for Arctic Citizenship analytical models created to assess levels of adherence to Sustainable Development (SD) and energy literacy. The integration of different theoretical approaches like Complexity Theory, Capability Theory, and Global Citizenship Education provide the theoretical foundations and contribute to enhance understanding about the nature of the 'energy curriculum' for Arctic Higher Education (HE) by providing data on the levels of energy literacy and on the adherence to Education for Sustainable Development. This study uncovers these educational elements and shows that the energy transition period will require well-equipped citizens engaged in a sustainable energy vision, a holistic approach to deal with several interconnected challenges materialized by the SDGs that are all interdependent of energy and education.

Natural processes and ecosystem services have contributed to economies and human well-being in a wide range of ways along centuries. The way natural resources are managed, among them, energy, exerts a fundamental impact on environment and societies at a geographical basis. The nature and features of energy systems can have important effects on how societies operate and contribute to Sustainable Development. Local energy systems also have a fundamental impact on education systems when considering mindsets and power structures aligned to non-renewable or renewable energy resources. For this reason, energy literacy is essential to open the spectrum of understanding in relation to interlinked matters like climate change, poverty alleviation, food security, potential conflict of interests (political and economic power, rights of indigenous people), energy security and (over) dependence on non-renewable energies, and how energy and human well-being are interconnected. Along the research project discussed in this book, it remains clear that energy education is a relevant component of sustainable energy literacy and an important path for shaping a sustainable society and a sustainable future, but it is also fundamental to be able to align knowledge systems, policies, and practices to a low-carbon economy. This is a starting point offering educational opportunities interlinked to several other components materialized by the 17 Sustainable Development Goals (SDGs) in order to empower the most vulnerable in society to adapt and live according to more equitable patterns of existence. Important commitments to make fundamental changes in the way societies produce and consume goods, commitments towards quality education, energy for all, sustainable management of natural resources among other commitments were established by this international framework. Based on the successes and lessons learned from the Millennium Development Goals, the United Nations, in 2015, started to focus on a more expanded set of 17 goals that perform the so-called Agenda 2030 (United Nations 2015b). The Sustainable Development Goals (SDGs) comprise a framework of ambitious global goals and targets designed to tackle the world's most important challenges as poverty, climate change, and inequalities but under the focus of SD. These 17 goals are as follows:

1. End poverty in all its forms everywhere
2. End hunger, achieve food security and improved nutrition, and promote sustainable agriculture
3. Ensure healthy lives and promote well-being for all at all ages

4. Ensure inclusive and equitable quality education and promote lifelong learning opportunities for all
5. Achieve gender equality and empower all women and girls
6. Ensure availability and sustainable management of water and sanitation for all
7. Ensure access to affordable, reliable, sustainable, and modern energy for all
8. Promote sustained, inclusive and sustainable economic growth, full and productive employment, and decent work for all
9. Build resilient infrastructure, promote inclusive and sustainable industrialization, and foster innovation
10. Reduce inequality within and among countries
11. Make cities and human settlements inclusive, safe, resilient, and sustainable
12. Ensure sustainable consumption and production patterns
13. Take urgent action to combat climate change and its impacts
14. Conserve and sustainable use the oceans, seas, and marine resources for Sustainable Development
15. Protect, restore, and promote sustainable use of terrestrial ecosystems, sustainably manage forests, combat desertification, and halt and reverse land degradation and halt biodiversity loss
16. Promote peaceful and inclusive societies for Sustainable Development, provide access to justice for all, and build effective, accountable, and inclusive institutions at all levels
17. Strengthen the means of implementation and revitalize the global partnership for Sustainable Development

The 17 goals and 169 targets of the SDGs were shaped as an integrated and indivisible framework to enable a global Sustainable Development (United Nations 2015a). They contain general guidelines unique in scope (United Nations 2015a), but this international approach has also been contested and revealed to be controversial as being considered a top-down strategy only focused on resolving global, regional, and local problems on the level of governments and international organizations (Hajer et al. 2015; Sköld et al. 2018) with a lack of customization to specific bioregions.

The importance of the SDGs for the Arctic is even more potentialized because a changing Arctic produces impacts on a global scale for the Arctic and non-Arctic world. The SDGs represent building blocks (Ganapin 2018) for socio-environmental transformation, but the most important characteristics is that they represent the shift from a fragmented thinking approach to an integrated thinking approach to SD. When education aligns to the SDGs, the same effect can be created in students and educators, in other words, a more balanced and integrated view and knowledge of the interconnections of fundamental aspects of sustainability that, so far, have been treated separately or in a silos style. It is the opportunity to create a proper SD system of practical solutions considering that everybody must be part of and have a voice in sustainable future. This is an important step to decolonize the mindset and the curriculum. At this stage, societies would be operating beyond the level of co-management for a prosperous future (Arruda 2018, p. 107) but towards a shift in the mindset to be an Arctic citizen towards an engagement to become a global citizen.

Perhaps the most important feature of the SDGs is that they break out of a siloed approach to development. Instead of pursuing social, economic, and environmental objectives in isolation, the Goals propose an integrated approach of interdependent and multidisciplinary components that should be considered under multicultural viewpoint. The SDGs, if known, understood and practiced, can provide a powerful local framework by supporting coherent, localized, inclusive, and sustainable decision-making. For this reason, it is so important to disseminate, embed, and measure the alignments and adherence of education and practices to the SDGs. However, to build up this data, there is the need for new regional and local mechanisms to inform people about them and to monitor how the SDGs are contributing to Sustainable Development in practice. Unfortunately, it is something that does not yet exist or is an on-going process in developed countries. In terms of the developing world – we can consider the Arctic as a developing region – the global Sustainable Development frameworks are not still a priority (Koivurova 2018) and this fact has also been recognized by the Inuit Circumpolar Council when expressing concerns about the inaction of governments of the Arctic countries towards the implementation of the 17 SDGs and their 169 targets (Eegeesiak 2018).

In reality, the commitment to the SDGs must be on the part of governments, businesses, institutions, and citizens in a true spirit of responsibility and Arctic Citizenship, accepting that multicultural stakeholders with different perspectives are co-creators and co-managers of knowledge, geocapabilities, and curricula involving multicultural perspectives and foundations towards adaptability to complex dynamic systems. Therefore, education plays a fundamental role in this process; however, if education is not committed to an integrated approach and aligned to SDGs or does not assess levels of adherence to the SDGs through ESD, the implementation of goals become abstract and difficult to achieve. This assessment will require the creation of mechanisms to evaluate how the SDGs can promote Sustainable Development in the Arctic. If there is a serious commitment for the SDGs to really advance Sustainable Development, it is important to ensure they reflect the Arctic's unique conditions and that it can be measured by clear indexes. Not only the encounter of TK and Scientific knowledge is relevant in this regard, but the encounter of the SDGs and TK is of extreme importance for realizing SD. How to incorporate this decolonized viewpoint into the SDGs will be probably the real challenge for the future achievement of the SDGs, because everybody needs to be part and to have a voice in terms of Sustainable Development as inhabitants of the Earth. This is again a case for this important humanity's shift in the mindset for feeling the genuine Arctic Citizenship responsibility and the global citizenship aspiration. When this personal mindset evolution happens, societies will be closer to SDGs achievement under an inclusive and participative reality. When these different players are part of a transformational dynamics, societies will be closer to perform and realize the SDGs.

9.3 Achieving the SDGs through ESD

In the Arctic, traditional livelihoods transcend the mere economic importance to be culturally and spiritually central to the identity and continuation of its peoples. The ways specific indicators will be developed to reflect these aspects will be of crucial importance to advance the SDGs in the Arctic.

One concerning aspect discussed in this chapter is the fact that no interviewee mentioned the SDGs when talking about their understanding of SD or when asked about ESD. The latest international surveys on the SDGs revealed evidence showing the lack of awareness and knowledge on the SDGs not only in the Arctic but worldwide, probably because education systems are not yet completely aligned to the 17 goals, as it has been concluded by the levels of energy literacy, or have not considered the multidisciplinary and multicultural perspectives of a sustainable vision in practice. Depending on the geographical area, this is not a disseminated subject. Global surveys like Eurobarometer and Globescan revealed that awareness of the SDGs is, in general, currently higher nowadays than awareness of the Millennium Development Goals was in previous surveys (European Commission 2018); however, according to the latest Eurobarometer (2017), just over 1 in 10 Europeans know what the SDGs and their geographical features are (Globescan Radar 2016; Special Eurobarometer 445 2017) (see Figure 9.1).

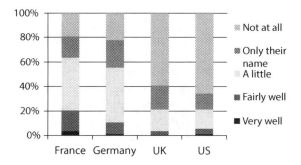

Figure 9.1 How well do you know the Sustainable Development Goals?
Source: OECD and Hudson and vanHeerde-Hudson (DevCom 2017)

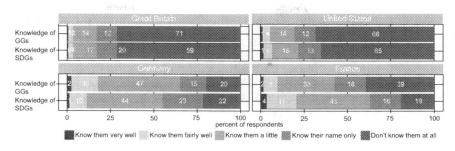

Figure 9.2 KNOW: Knowledge of Global Goals and Sustainable Development Goals. Source: Globescan Radar (2017); nationally representative samples of approx. 1,000 adults in 16 countries (Fieldwork: 12/16-05/2017).

To understand how these processes interconnect and how to decrease local vulnerability due to environmental change, it is crucial to promote cross-fertilization among a diversity of knowledge systems that can contribute with new evidence and also improve the capacity to interpret conditions, and response to change. This is a viable way of making indigenous communities part of the adaptive solution under a systematic approach to coordinate different SDGs related to very specific bio-regional economic activities related to food security, like reindeer herding, fishing, hunting, and at the same time promoting adaptation to cultural practices and traditional economic systems that represent local livelihoods.

Different knowledge systems are made up of agents, practices, and institutions that organize the production, transfer, and use of knowledge (Cornell et al. 2013). To engage different knowledge systems means to make all these stakeholders part of a Sustainable Development process considering the intergenerational dimension of SD. It seems to be an important path of inclusiveness for the future of Sustainable Development.

Apart from the main issues and indicators related to the Goal 4 (Education), there is a significant lack of information and knowledge in relation to the SDGs all over the world. The importance of the SDGs to the Arctic is enormous and difficult to narrate in words due to the complexity of its myriad systems. There is the need to apply mechanisms of measurement in relation to SD adherence, levels of SDGs dissemination, and SDGs adherence in difference arenas. One fundamental tool is HE aligned and oriented to SD via SDGs in order to reach as many people as possible and provide ways for them to be part of the debate and the daily application of the goals and targets of Agenda 2030 (UN-United Nations 2015b). This requires a sophisticated level of embedded SD that will not be achieved without a systematic and inclusive process of Education for Sustainable Development (ESD) to form Arctic citizens and global citizens. Countries and institutions seeking to achieve the SDGs need to systematically disseminate the SDGs, listen to their citizens, to engage with them, to promote policies of inclusive participation and mobilization towards action on the goals. Understanding what different stakeholders know and think about the SDGs and developing data sources to shape indicator frameworks to follow up the progress towards the goals through national strategies and statistical systems would also contribute to SDGs achievement.

Higher Education for Sustainable Development is an incipient but an important branch of HE, because education is both a goal and a means for attaining all the other SDGs. 'Education can, and must, contribute to a new vision of sustainable global development' (UNESCO 2017, p. 7). ESD is the path for SD and a key strategic mechanism in the pursuit of the SDGs. SD is still lacking a systematic integration and reflection in the mainstream of HE. In this sense, education policies should promote HE alignment to the SDGs via a culturally inclusive curriculum, multicultural contents, approaches, pedagogies, and systematic curriculum design and review oriented to SDGs learning objectives (UNESCO 2017). ESD can be a key educational tool with profound impacts on learning when developing specific bioregional learning outcomes needed to work on achieving particular SDGs (UNESCO 2017, p. 10). The realization of a true Higher Education oriented to Sustainable Development with successful achievements in SDGs arena will depend on these goals and targets to become the daily practice in communities based on the recognition of rights and culture, the exercise of values, the development of skills, and geocapabilities in people that to know how to live together and work towards an inclusive and balanced development model by respecting the support capacity of the planet. It will require a deep transformation in systems and technology but primarily in nations and in citizens.

References

Arruda, GM 2018, 'Artic resource development. A sustainable prosperity project of co-management', in GM Arruda (ed.), *Renewable energy for the arctic: new perspectives*, Routledge, Abingdon, p. 112.

Cornell, S, Berkhout, F, Tuinstra, W, Tàbara, JD, Jäger, J, Chabay, I, de Wit, B, et al. 2013, 'Opening up knowledge systems for better responses to global environmental change', *Environmental Science & Policy*, vol. 28, p. 60.

DevCom-OECD Development Communication Network 2017, '*What people know and think about the sustainable development goals*', Selected Findings from Public Opinion Surveys. OECD Development Communication Network, November, Paris.

Eegeesiak, JO 2018, 'Are the UN sustainable development goals being achieved in Inuit Nunaat?', *The Circle*, vol. 2, pp. 26–28.

Eurobarometer 2017, New European parliament, www.europarl.europa.eu/news/en/headlines/priorities/eurobarometer-2017.

European Commission 2018, *Education and training monitor*, Brussels, Belgium.

Ganapin, D 2018, 'The UN SDGs: providing the building blocks for a more sustainable future', *The Circle*, vol. 2, pp. 3–28.

Globescan Radar 2016, Nationally representative samples of approx. 1,000 adults in 25 countries, (Fieldwork: 12/15–05/2016).

Hajer, M, Nilsson, M, Raworth, K, Bakker, P, Berkhout, F, de Boer, Y, Rockström, Y, et al. 2015, 'Beyond cockpitism. Four insights to enhance the transformative potential of 139 the sustainable development goals', *Sustainability*, vol. 7, no. 2, pp. 1651–1660.

Koivurova, T 2018, 'Why the arctic needs the UN sustainable development goals', *The Circle*, vol. 2, pp. 6–28.

Simkiss, D 2015, 'The millennium development goals are dead; long live the sustainable development goals', *Journal of Tropical Pediatrics*, vol. 61, pp. 235–237.

Sköld, P, Liggett, D, Smieszek, M, Staffansson, J, Bastmeijer, CJ, Evengård, B, Muir, M, Scheepstra, AJM & Latola, K 2018, *The road to the desired states of social. Ecological systems in the polar regions*, EU-PolarNet White Paper.

Special Eurobarometer 445 2017, 27,929 individuals surveyed across 28 EU Member States, (Fieldwork: 11–12/2016).

UNESCO 2017, *Education for sustainable development goals: learning objectives*, UNESCO, United Nations Educational, Scientific and Cultural Organization, Paris, France, pp. 7, 10.

UN-United Nations 2015a, *Millennium development goals report*, United Nations, New York.

UN-United Nations 2015b, *Transforming our world: the 2030 agenda for sustainable development*, Resolution adopted by the General Assembly on 25 September 2015. Document A/RES/70/1, United Nations, New York.

10 Multidisciplinary and multicultural perspectives of a sustainable energy system in Arctic Higher Education curricula

Analysis, discussion, and conclusion

10.1 Introduction

This concluding chapter aims at analysing and discussing the findings (from interviews and documents) and it is oriented to address the central aim of this book that is to investigate if the multidisciplinary and multicultural perspectives of a sustainable energy model are being addressed by HE curricula in the Arctic.

This chapter presents a discussion commencing with the indigenous and non-indigenous contexts of Arctic Higher Education exploring features of indigenous education and the vocational vs. academic dilemma of the non-indigenous style of education. This chapter highlights the more significant elements that emerged from the engagement in Phase 1 (Documents), Phase 2 (Data Collection), and Phase 3 (Data Analysis) of the research project.

Some tensions and controversial reports on the SD and energy literacy concepts emerged from the interviews' data in terms of indigenous students' perspectives about SD and the controversial feature of energy (production and use) in some specific Arctic areas. These tensions provide evidence of the necessity to enhance understanding of the multicultural dimension of SD and ESD as a tool for Arctic adaptation and the importance of energy literacy in Arctic Higher Education curricula.

In this book, energy has been used as a lens to focus on the sustainability debate. The critical element of change analysed in the Arctic was energy production/use and energy transition in sensitive areas impacting human and natural systems as well as the capacity to understand rapid and complex geographical changes, and benefit from a multicultural dimension through HE. The complexity around sustainability serves as an educational impulse for the improvement of learning processes as exposed in Chapter 2. Ryan (2011, pp. 3, 5) argues that the continuous search for sustainability presents an opportunity to develop education and pedagogies. This book goes beyond this point to argue that the current sustainability challenges depend on a deep and meaningful understanding of the concept of SD and its components, if the concept is operative or lost the meaning and what it entails to make it practical for all in different Arctic contexts. According to findings presented in previous chapters, SD means different things to different people. Different groups show different expectations from SD concept, principles, theories, and practices. These expectations are slightly different when distinct cultural and economic (determined by distinct energy systems) backgrounds are considered. While the concept is broad for some people, for others it lost the meaning. For some of the interviewees (lecturers and students), SD does not have an operative meaning into practice, being only a political jargon, which makes it a temporary instance or an impractical reality in need of review.

SD is a contested concept in the literature, as the findings of this research have evidenced and authors like Baker et al. (1997) have argued. The concept is controversial even when non-indigenous stakeholders from different energy backgrounds (oil and gas, and renewable energy systems) are considered, and this lack of convergence is even more emphasized when indigenous and non-indigenous staff and students' perspectives are under analysis. Findings from staff and students' interviews also corroborated the argument of Hove (2004, p. 53) that SD is a catch-phrase or jargon as the economic pillar continue to prevail over environmental and social concerns (Fernando 2003, p. 31) representing a contradictory approach to mediate the impasse of development. In the literature and in practice, the balance among the TBL pillars seems to be difficult to achieve.

ESD is thus also essentially contested territory and open to various interpretations following the same controversies of the SD concept and not an educational mainstream because of the same difficulty of providing the practical balance between economic, environmental, and human aspects. ESD is a vision of education based on important premises of compatibility of the human needs to the natural capacity of the planet and solidarity inspiring the notion of citizenship, but these are premises that vary according to the cultural and economic backgrounds. In the Arctic Higher Education staff interviews raised a variety of conceptions of SD and purposes of ESD were shown depending on the style of education. Likewise, on the students' side, there was also a variety of ideas about these two concepts. The dominant factor, however, was the indigenous vs. non-indigenous conception, which aligned well to a practical vs. theoretical and life-long learning vs. learning for employability dialectical approach. As a way forward to integrate the best of both, a more cultural dimension can be acknowledged and, as a result, a citizenship framework has been proposed to develop existing models presented in Chapter 3.

The pursuit of SD represents a remarkable educational opportunity, but the concept needs further understanding, expansion, and a balanced practical meaning. A deeper understanding is essential to motivate changes in the education system to make it more inclusive, comprehensive, and dynamic. Different perspectives should be considered when discussing SD in order to inform better decision-making in an exercise of inclusivism and democracy.

10.2 Contexts of Arctic Higher Education – indigenous vs. non-indigenous education

10.2.1 Indigenous perspective and education

The Arctic presents a natural stage for intercultural learning. The vast distances and remote locations where small communities were formed is a factor that stimulates multicultural encounters in the academic environment and through on-line platforms by creating opportunities for learning with equally productive intercultural encounters. Contemporary information technology has revealed to be an important facilitator to multicultural dialogue. Indigenous communities are no longer understood in solely geographical terms. Although new information technologies may reinforce the existing structures of power (creating a digital divide), indigenous groups can use the same technologies to their advantage and challenge the mainstream representation and narratives (Arruda & Krutkowski 2017a, p. 519).

In Chapter 2, it has been provided an overview of the dimension of indigenous presence in the Arctic showing the importance of rethinking Arctic Higher Education and the dimension of the sustainability and natural resources (energy) debate through the contributions from the possible exchanges between indigenous and non-indigenous HE curriculum for the Arctic. According to Ascher (2002) and Eglash (2002) indigenous people have had their own ways of looking at and relating to the world and they have demonstrated a relative lack of enthusiasm for the experience of traditional Western-style education due to cultural reasons and the institutionalized forms of cultural transmission (Barnhardt & Kawagley 1999).

The First Nations education, for instance, has been involved in political and racial construction of indigeneity, truth, and reconciliation, and the evolving social construction over time (Atleo & Fitznor 2010) has created a level of tension that has been reflected in HE. Indigenous and non-indigenous people in the academy have realized the limitations of a mono-cultural education system, mainly when controversial exploration projects (energy and minerals) are implemented in their lands as sustainable and beneficial to the locals. Indigenous educators and students realize the importance of learning about science (chemistry, biology, physics, energy, and scientific theories) (see Section 8.2, Florence's reports), while non-indigenous scientists/academics started valuing TK in science and education in adaptive initiatives like EALAT and ELOKA (McCann, 2013) confirming positions of previous research by Barnhardt and Kawagley (2005, p. 8):

> Recently, many Indigenous as well as non-Indigenous people have begun to recognize the limitations of a mono-cultural education system, and new approaches have begun to emerge that are contributing to our understanding of the relationship between indigenous ways of knowing and those associated with Western society and formal

education. Our challenge now is to devise a system of education for all people that respects the epistemological and pedagogical foundations provided by both indigenous and Western cultural traditions.

While Western science and education tend to emphasize compartmentalized knowledge, which is often de-contextualized and taught in the detached setting of a classroom or laboratory, indigenous people have traditionally acquired their knowledge through direct experience in the natural world. For them, the particulars come to be understood in relation to the whole, and the 'laws' are continually tested in the context of everyday survival.

(Barnhardt & Kawagley 2005, p. 8).

Arctic Higher Education offers an invaluable opportunity to this multicultural approach in education and it is a region naturally oriented to problem-based and active learning that prepares students not only for the workplace but also for life challenges. Competencies to deal with professional and personal challenges need to be part of Arctic Higher Education because the Arctic presents a real-life learning setting requiring life-long learning approaches, and geo-capabilities for prompt application of knowledge and skills in the field.

Another essential characteristic is that learning does not only happen individually but collectively. Collaboration and community engagement (another indigenous cultural characteristic) represent invaluable opportunities for learning at all levels and for showing a sense of global citizenship, because life on ice is an extraordinary extra-curricular activity that brings important practical learning experiences to the HE curriculum.

The discussion of fundamental topics like SD and energy were more relevant where the academic environment was indigenous or mixed for comparative reasons. In this regard, representative cases were from Glacial University (GU), Juniper University (JU), and Ice University (IU) where staff and students (interviewees) were from different indigenous groups, including Inuits, Metis, Crees, Sámis, Inuus, Denes, Inupiat, Yupik, and Aleut.

The findings revealed an important asymmetry in the understanding of the concept of SD between indigenous and non-indigenous participants. SD means different things to different people from different economic backgrounds and from different cultural knowledge constructions. Lecturers emphasized the idea that it is a difficult term to define; consequently, it is a difficult term to explain and teach about in class. Different people perceive the concept of SD in different ways according to their own economic and cultural systems that provide different perspectives and reactions towards it. The consequence is that it triggers different processes of decision-making in terms of designing curricula, with different outcomes. Varied learner motivations and expectations also bring nuanced perspectives to learning 'about, for or through' SD (Sterling 2003).

Apart from the apparent difficulty associated with the concept, lecturers were explicit in confirming the importance of promoting understanding of the concept as a primary objective of their courses. For indigenous lecturers the understanding of SD is not the same as in Westernized education. SD means the continuation of an indigenous culture and livelihoods in an economically consistent way. These lecturers were clear about the fact that 'not everybody understands the meaning of SD because the concept lost the meaning along the time, being just a political phrase'. This was also a position endorsed by indigenous students that emphasized the lack of practical sense of the concept.

Indigenous lecturers identified a difference in conceptual understanding of SD in relation to the Western academy because the concept relates to a longer period of time that is not comprehensible to the human being's mind, but it is understood in indigenous systems (Section 4.2.1.1) and linked to intergenerational survival of the group. The intergenerational dimension is a characteristic of indigenous culture, and a meaningful concern demonstrated by the staff and students' reports of different indigenous nations represented in the sampling of this research.

The relevance of creating awareness about the intergenerational socio-environmental impacts and economic effects of those impacts seem to guide lecturers to reflect on collective responsibility of developing, in society, alternative approaches for community engagement into learning, the student's role in SD and the students' role towards a sustainable world. This approach to awareness of a long-term development view can also be explained by some indigenous opinions (staff and students from different locations) in relation to the concept's usage as politicians employed SD to short-term political campaigns and mandates (four year-period) making the term to sound as electoral jargon. This is the feeling of some interviewees that reported living in areas they consider completely unsustainable as food and fuel (fossil energy) come from outside at huge costs for the locals and for the environment. It is comprehensible that for those living unsustainable daily experiences it is a matter of not only understanding the concept but also experiencing hope to change the system.

Indigenous lecturers understand the concept of SD through an integrative approach (multidisciplinary approach) by discussing geo-bio-physical, socio-environmental and climatic effects and linking them to knowledge creation in the benefit of communities. They recognize the importance of educating the youth about these changes by establishing their role in the process of promoting knowledge creation, dissemination, and adequate local adaptation to the on-going changes through the nurture of multiple skills. Indigenous educators focus on society, and on local communities, looking at issues under local, regional, and global perspectives (holistic viewpoint). They see ESD as a tool to empower indigenous youth to bring their livelihoods and culture to the future as Grete and Eva (indigenous lecturers) emphasized several times in their interview reports (see Sections 4.2.1.1 and 4.2.1.2) corroborating positions of Orr (1992), Poppel (2015) and Magni (2016) on the importance of culture to SD.

In indigenous education, sustainability does not happen in an educational setting, but in the real world, the Arctic world. Indigenous lecturers and students justified this position saying that SD is a concept that varies according to context and culture. Sustainability (as a function of SD) is considered a cultural value or, in other words, it is defined by culture. According to the indigenous perspective, values are taught by doing not only by rhetoric, they are taught by example, by application, by practice confirming the dynamicity of TK argued by Sillitoe (1998). Kristine (indigenous lecturer; see Section 4.2.1.2) was decisive when saying that it is not possible to inculcate values and behaviours that have been transmitted by generations and family tradition through formal education and teaching; she advocates the position that values can only be taught by practice and are difficult to incorporate in a lecture-base Westernized style education as rhetoric is not easily assimilated by indigenous students. Indigenous students have different ways of learning.

Another fundamental point made by indigenous lecturers and students relates to the difficulty in the Western education system to incorporate more than one perspective in formal discourse and education. According to interview findings, there is a mismatch or a cultural clash impacting teaching and curriculum design. Indigenous lecturers believe that there is the need for the students to be taught different viewpoints, as the current educational system just considers one viewpoint (the viewpoint of the colonization process). Some indigenous students illustrated this argument by reporting that in some northern Westernized universities, lecturers teach that indigenous economic activities are harmful to the environment and it can create tensions when there are indigenous students in the group. The tendency in the literature has been to approach contexts from a non-indigenous perspective, as the major part of academic bibliography has been written with focus on a Western/scientific view of the world (Orvik & Barnhardt 1974; Smith 1999). However, there has been a debate about the value of multicultural perspectives in teaching and curriculum design involving the encounter of Western science education to what local people know and value because there are native ways of knowing (Stephens 2000, p. 10) and the recent trend is to envisage this encounter in parallel to the sustainability debate (Arruda 2018; Bitz et al. 2016), because exploration of a topic through multiple knowledge systems can only enrich perspective and create thoughtful dialog (Stephens 2000, p. 10) and the modern Arctic education requires knowledge and information to be created and disseminated in a variety of formats intelligible to a variety of stakeholders.

In terms of natural resource management, the indigenous perspective tends to consider natural resources, as an interconnected system. There is a more integrated (holistic) view (not compartmented in disciplines) on how to manage them as water, land, energy, livestock, and people are part of the same landscape of multiple interactions. It is not a picture; it is a dynamic and holistic reality of interconnected elements moving in the opposite direction of a reductionist worldview (Boulton et al. 2015, p. 66; Wheatley 1999, p. 10; Youngblood 1997, p. 27, p. 10) based only on the economic pillar. TK is a type of knowledge in constant evolution that tends to break the chains of cause-and-effect models towards adaptation, continuation, and pragmatism.

Based on findings, this book argues that the study of natural resources is not done by resource individually considered but taking the interactions among them and the connection with biological processes. This is one of the reasons why indigenous educators study SD through the prism of biological diversity, because diversity means life and its myriad of complex interactions. SD for indigenous groups means to conserve life. The other reason is alignment to the Convention on Biological Diversity (CBD) with the most important indigenous rights charts. Additionally, managing natural resources consists of developing specific skills of assessing the natural assets and managing them in a prudent way to sustain community's livelihoods of the present and the future. Resource management is seen as an essential skill important to everyone not only for the learner. It is considered a skill to be understood, acquired, and applied by everyone in the field. Students are stimulated to enhance knowledge about the impacts of extractive activities on the environment, to debate about economic activities and their socio-environmental impacts as well as to apply this skill to create solutions to real-life problems. The application of knowledge is of paramount importance in native cultures (Ilutsik 2000, p. 30). These nuances offer a special perspective about SD to the Western academy.

In the indigenous perspective, SD implies the continuation of indigenous culture and livelihood. It means to have available natural resources for the present and future generations to be able to use the resources the land provides and to be able to survive. This skill is exercised in the real world and it is not only nurtured in the academy, but also in the community. The management of natural resources and the intergenerational dimension are part of the culture; they are not seen as standard provisions of SD concept or international treaties. Communities are the end users of knowledge created in the university and the courses have a theoretical component, but practice is the purpose of study and the learning experience for indigenous students and lecturers. Intergenerational survival and development of communities, and the continuation of indigenous culture and livelihoods are the main objectives of the indigenous learning experience and learning environment, approaches and pedagogies. Indigenous students are co-creators and disseminators of knowledge in the communities. In this sense, knowledge has an impact in the real world. This active knowledge is the expression of what is to be indigenous (Nussbaum 2000, p. 72) (Sen 1993, p. 31); it is the expression of power and freedom (Freire 2000, p. 50). This book therefore concludes that it is not possible to provide welfare and freedom without understanding what is sustainable from different perspectives and exercising the concept in practice. Both indigenous and nonindigenous educational systems have creative branches and elements to be exchanged and improved, however, in terms of SD the indigenous viewpoint demonstrated to be more pragmatic, more community-focused and connected to capacities and skills of temporal continuation of the group (intergenerational dimension) what makes them to operate within natural resource limits. There is a natural attachment to place that is considered home, and a sense of care and connection to all forms of life in this place called home (Arruda & Krutkowski 2017b, p. 279). These elements are meaningful to the indigenous experience of the everyday life and unnegotiable if a decision about an unsustainable project/practice needs to be made within the community.

In general, elderly members of a community are considered the knowledge keepers of TK that is transmitted by art, language, history, music, medicine, etc (Battiste & Hernderson 2000). For an indigenous student, these are the primary sources of TK and an innate mechanism of learning. In an indigenous educational institution like the Glacial University (GU), indigenous students debate and analyse contexts under a cultural and holistic perspective. The course looks at themes from an indigenous viewpoint and follows an indigenous methodology and epistemology. The course objectives focus on learning about the relations with people, environment, land, and ideas, and education takes place by considering cultural values, challenges, and how to promote adaptation to change, complexity, and uncertainty in order to preserve the group. Change and adaptability are the essence of TK; for this reason, this book advocates the position of convergence between science/Western education and TK.

TK is a dynamic repository of information for the Sustainable Development and well-being of the communities through the generations. TK informs SD and natural resource management in a practical and active way. Knowledge and capabilities for indigenous people have a clear purpose to be produced and the purpose is the application to communities' daily life, for well-being, personal and collective development, intellectualism, reflection, actualization, social transformation, knowledge transmission, and intergeneration adaption.

Indigenous knowledge is open-minded. It is open 'to widen definitions of knowledge' as Grete (indigenous educator) confirms the words of Longman (2014, p. 17). This is a very particular objective of the course studied at Glacial University (GU). This objective means that knowledge is an open chapter; it is not static, it is not mechanic, but active, dynamic, receptive, and adaptive to how complexity, uncertainty, and change require knowledge to be. The consequence of this perspective is that SD is equally an open, active, dynamic, and adaptive notion to assure the intergenerational survival of the communities.

Education and learning follow the indigenous methodology and take place based on several levels of relationships, with people, with the environment, with land, and with ideas, meaning that the learning is practical and occurs as a blended experience between the academic and the world reality. TK is researched, compiled, and applied to be used as a source of information for the durable development and well-being of communities.

The co-creation of knowledge by students is accomplished within the indigenous context being originated from the primary data acquired from indigenous people, from the cultural engagement and participatory approach within the community through interviews, apprenticeships, storytelling, cultural observation, filming, and photographing, where students are researchers co-creating knowledge according to an indigenous epistemology. The documentation of TK is not only a way to protect cultural heritage but a form of transmission of knowledge for future generations, making knowledge a component of the intergenerational dimension. It means that the systematic documentation and use of TK contributes to intergenerational knowledge transmission. Knowledge means continuation and identity. Knowledge is managed through time and it means power.

To expand and disseminate communities' and societies' knowledge about biological diversity according to their local experiences is a central aspect of the teaching and learning at the Glacial University (GU). Knowing relates to a theoretical capacity, while a skill is linked to a practical ability. Learning how to document TK involves a process of documentation in which skills turn into knowledge according to Grete (indigenous educator). In other words, practices are systematized to turn into stories (theoretical, written material). The practice turns into 'theory'.

In this sense, knowledge is not theory, but it is originally practice; in a Western way of thinking, academics would argue that 'knowledge is theory into practice', but for an indigenous person (educator or student), these two dimensions co-exist at the same time (there is no 'theory' and 'practice'; they are not divided into parts (fragmented into disciplines) to be exercised in different moments, but at once. When students return to their local communities to document TK, they are trying to record and systematize their practices by using methods like audio recordings, drawings, interviews, filming, and photographing, as a means to capture specific practices of indigenous resource management systems and the basis for natural capital preservation. This integrated mind-set on knowledge and the clear purpose of co-creating knowledge are the real legacy of indigenous education according to this book because indigenous ways of learning are different and happen in a different way in space and time.

Some lecturers believe that this understanding of SD and energy is formed within activities of everyday life. These lecturers emphasised the land impacts caused by energy projects and the importance of discussing with the students about the responsible use of natural resources and land. Land is a crucial resource for Arctic indigenous communities and the study of the interactions contribute to enhance knowledge about the impacts of extractive activities on indigenous land that sustains people's livelihoods.

SD is embedded in the course despite not being labelled 'sustainable development' but, instead, biological diversity. When students go and document biological diversity through the practices adopted in his/her area and later share this experience in class this widens definitions of knowledge. Along the circumpolar Arctic, there is a myriad of communities with their ancient practices, which are shared, interpreted, and translated into the Arctic indigenous context. The shared experiences in class based on documented TK provide a common basis for understanding the region and the surrounding world.

SD is translated by conservation of biological diversity, in a way that the conservation of life in the physical environment (geography) is an intrinsic premise for the conservation of people's culture and livelihoods in relation to human environment (human/social geography). These are two indissociable elements, meaning that when a student researches the physical environment, he/she is also studying the human environment because the pillars of sustainability are seen as one (economic, social, and environmental). This book has argued that the cultural aspect has been under recognized (Section 2.7) through adoption of the traditional TBL model. An indigenous perspective has no pillars; it is just one thing because practical knowledge comes from people and from the environment with its myriad interactions; however, it is deeply cultural.

Documented TK has the purpose of being shared and accessible to communities to provide cultural continuity (intergenerational dimension) through social attitudes, beliefs, conventions of behaviour, and practice (Berkes 2008, p. 3). Due to the importance of TK, technological means have been used to record and store it by creating specific databases. These digital tools have had a significant impact on indigenous education and Arctic pilot projects in remote Arctic communities (EALAT and ELOKA). The range of profound changes seen in the Arctic region, however, motivated the use of satellite, digital technology, the internet, and social media to strengthen cultural and linguistic connections by storing and disseminating information in order to promote common interests. Indigenous people use media as a tool for cultural and political intervention, addressing information gaps, combating stereotypes, and reinforcing local community languages (Meadows 2009, p. 118). Through new media and technology, indigenous people can pursue community-focused goals, learn, and disseminate knowledge. They can exchange information, preserve oral traditions, connect dispersed communities (diasporic groups), communicate, and share resources to collectively tackle social issues affecting their well-being and identity (Srinivasan 2006). This extraordinary process of dialogue and interaction is ongoing across the Arctic nations like The First Nations (Innu, Dene, Crees), Inuits, Metis, and Sámis, who had representatives in the researched case studies as participants in the interviews, in Norway, Canada, and Alaska.

Indigenous versus non-indigenous education represents a 21st-century tension in need of being efficiently addressed by ESD. This is a central point for HE in the Arctic. ESD can contribute in the process of evolving a mutual understanding of SD based on multiple cultural frameworks if it privileges reflection on a truly multicultural perspective because each culture constructed worldviews based on its interpretation of the concept. The new ESD model presented in this book can help people understand the divergences from differing cultural frameworks and search for convergences (common denominators) based on ESD adherence parameters. What is proposed here as 'Type 4' or 'Arctic Citizenship Education' represents this desired multicultural understanding and level of SD literacy with no pedagogic imperialism. This multicultural understanding about the concept of SD could create a threshold for new educational multicultural spaces.

The creation of multicultural educational spaces where indigenous and non-indigenous feel safe and comfortable to be whom they are and express their opinions according to their personal and collective experience depend on the pragmatic implementation of multidisciplinary and multicultural perspectives in the HE curricula based on the best levels of SDL (in its fundamental component – energy literacy). These multicultural educational spaces depend on the application of a true Arctic citizenship and the capabilities and functionings, with all the attributes that it entails. These Arctic multicultural educational spaces will be more enriched by a better understanding of TK (knowledge and values) and the indigenous perspective oriented to a dynamic and in constant adaptation knowledge, its immediate application on real-world problems with the purpose of wellbeing representing the construction of new and alternative realities.

This Arctic multicultural educational space, proposed in this book, also requires multicultural teaching and research *praxis* and pedagogies. The indigenous ways of teaching and learning bring on board important contributions related to curriculum content, case-studies, systemic thinking, and real-world problem solving, where research is focused on cultural well-being and teaching techniques to deconstruct the ideologies of representation, stereotypes, and discrimination, where narrative and enquiry are widely used in class to privilege reflection, critique on literature, and meaningful engagement with the knowledge creation process as reported along the findings of this research that corroborates aspects raised by Tuhiwai Smith (1999, 2005). The encounter and gradual integration of scientific knowledge and TK has the potential of enriching not only the academic environment but also the real world towards a sustainable mind-set, because culturally knowledgeable students demonstrate an awareness and appreciation of the relationships and processes of interaction of all elements in the world around them. They can live together (Delors 1996, p. 37) and build healthy and ethical relationships with people from different backgrounds; they can understand the interconnections between the natural and social fields; they understand geography and the bioregion where they live; they co-relate different knowledge systems and can anticipate and adapt for change; they can be who they are and appreciate their position, value and contribution to the world. This is a process that tends to benefit all the parts involved.

10.2.2 Non-indigenous education: vocational vs. academic

An important dilemma clearly demonstrated by the interviews is the 'employability versus knowledge application dilemma'; it means that there is a concern on the side of lecturers in training students to be professionals, but there are no data whether the knowledge was really applied in career or in life. On the other hand, students are also concerned with getting the certificate in the right subject to have a job after their studies. Application of knowledge is an expected outcome that is not measured or documented. There is a 'good intention' and 'an expectation' that the knowledge is applied, but no one can say if it really was or will be applied into practice. Students can apply their knowledge of SD and energy when returning to their countries, which theoretically would represent knowledge application into practice at a local level (according to geographical characteristics) and a positive impact of the course translated into geographical customized energy and development solutions. Apart from official and funded joint projects developed with other institutions, it seems not to be possible to measure the individual application of knowledge after the course when students return to their home countries. This is possibly a relevant aspect to include in the curriculum design and review – that the courses are oriented to practice and evaluated by alumni.

The discourse in the Western-style academy recognizes society as valuing 'informed-educated knowledgeable citizens', but it is not clear what it really means to be a citizen and what the role is of these citizens in this perspective. It is also not clear if the concept of 'global citizenship' (Delors 1996, p. 37) is really a meaningful, understood, and practical definition because the trend verified at Borealis University (BU), despite the students' advanced levels of awareness and interest in SD, was that knowledge and skills tend to have the purpose of equipping students for future employment. Employability and funding are the relevant purposes of acquiring knowledge and skills, but not enough to develop geographical consciousness and citizenship.

The combination of geography and business strategy in relation to natural resources is another characteristic of Western-style Arctic Higher Education promoting an understanding of energy dimensions (business, government, economics, policy, geography, finance risk, etc.) in order to prepare students for the energy industry, by training future energy industry professionals. These specific courses focus on preparing students to operate in the energy industry as energy workforce. Courses aimed at workforce development do not seem to be open to a more cultural citizenship approach to ESD according to the sample of this research. These are courses that focus on technical aspects of energy extraction, economic performance, environmental impacts, and how to address these impacts by using technology and environmental management to face the growing global energy demand. The idea of developing a new and alternative curriculum that was holistic, inclusive of indigenous world views (valuing TK), and prepared students for the workforce is probably a good aspect for future research amplifying the number of Arctic educational institutions to be analysed, but, fundamentally, it will only be possible based on a clear understanding of the multicultural and practical components of SD and based on the premise of a local sustainable energy system mind-set (renewable energy). Students under this new curriculum will not only be prepared to work for a power plant, but they will be prepared to work everywhere across the Arctic, in urban and in remote communities' environments.

10.3 The nature of the 'energy curriculum' within sustainability-related programmes to evaluate embedded levels of energy literacy

In Chapter 4, the first findings were presented according to the lecturers' reports and were consolidated in a code table and in charts to remark on the main features of the course objectives at the four case study institutions. This was a way of visualizing the most relevant characteristics of the course objectives and providing a comparative and more systemic analysis of the findings. It is important to remark that despite the six patterns verified by the coding the components have different meanings (practical/theoretical) at the four institutions under study.

The four courses presented common patterns related to the relevance of helping students *to understand the concept of SD* (and within this concept the fundamental role of the energy component). It is also relevant to realize that the conceptualization varies according to the geographical context, and the emphasis given by the lecturers shows different components of the concept according to economic and cultural alignments. The courses present a common concern that is to help students to understand the concept of SD but the approaches to achieve this objective are different according to cultural and economic local characteristics. The concept of SD can be approached by the prism of biological diversity, or by the prism of the WCED (1987, p. 43) 'Our Common Future'. The second common pattern verified among the lecturers' reports was concerned with *awareness – raising*. At first, this code seemed a bit abstract (theoretical), however awareness was promoted in different ways. It was reported that courses had the objective of making students aware of SD and energy issues. Despite the commonality of this pattern among the four case studies, the manner through which courses operationalize this objective varies according to the economic and cultural contexts and perspectives. Awareness can be cultivated by practice, e.g. cultural meaningful experiences in the field (in indigenous education) or it can be motivated by facts, arguments, and evidence in class based on theories and international agreements.

The third pattern related to the course objectives was '*to understand the interactions among natural resources, energy, environment, and society*'. To study SD only through ecology or only through climate change is to think under a limited scale. To solve large-scale problems, require an amplitude of interfaces of everyday life factors and to develop a sense of responsibility in dealing with them. These are interactions that indigenous peoples have more familiarity with as indigenous ways of life and well-being naturally depend on these interactions. This is not an aspect that indigenous students need to be taught in class, as staff and students raised at Borealis University (BU). To understand connections among systems (natural and social) is a cultural characteristic innate to the indigenous way of life and TK corroborating aspects argued by McCann (2013).

The fourth code was 'to make students aware of natural resources management'. It is important to analyse if the courses are equipping students for jobs or equipping students to be co-creators, disseminators, and appliers of natural resources management knowledge and skills in the field. Again, here the words seem to be the same, but the experience has different meanings. To assume that the problems humanity faces are essentially environmental and can be resolved by technology and natural resources or energy management (Scott and Gough 2003, pp. 113, 116) means learning about SD (Type 1) not living SD and this book argues that it is a deficient educational approach for the Arctic reality. Natural resources management can also have completely different focus and practical meaning if oil and gas extraction or renewable energy projects are being developed in the studied locations because the first type of resource implies the management of impacts on health and environment through technological solutions that are not guaranteed to prevent hazards, and the second case implies a different focus oriented to well-being and durability.

The fifth common pattern among the courses was related to intergenerational impacts. This is a matter of assessing how meaningful the expression is in practice in the different institutions. It can mean the continuation of communities in the long run, or it can mean a provision of an international treaty. This specific code gives evidence of the contrast between indigenous and non-indigenous education and the contribution TK can give to Western HE. Intergenerational dimension is a cultural characteristic to the indigenous groups studied as they have a natural concern in perpetuating the group as it has been done in the last 2000 years. For the Western-style education, it is not an innate component of the everyday life, but it is discussed as a provision of the SD official international concept. It has different practical implications in terms of an effective understanding of SD concept and, consequently, in relation to behaviour change, action and decision-making.

232 *Multidisciplinary and multicultural perspectives*

The sixth code or *critical thinking* is a common pattern among the four case study universities and represent a positive alignment of the Arctic HE system. All the courses aim at developing student's capacity of 'critique' according to similar and distinct approaches. In an indigenous academic environment, critical thinking is developed by using indigenous and Western scientific literature, via an inquisitive method to stimulate thinking and the formation of personal opinions on topics and practical experiences. In a Westernized style of education, critical thinking is essentially stimulated by literature, debates, and in-class activities and has a critical role in creating a sustainable future as per Cortese (2003). Students from the four case study universities emphasized the importance of critical thinking in their learning experience.

Despite the fundamental importance of these codes (course components), they are insufficient to address the contemporary Arctic challenges faced in education institutions as in the field. Arctic energy geographies require a more systematic thinking reflected in the curriculum under a more multicultural, multi-perspective, holistic, 'Arctic citizenship' progressively oriented to global citizenship. There is not enough emphasis on essential concepts like energy efficiency that present a vast opportunity for the Arctic (The Energy Efficiency Partnership 2013). It is important to demystify energy literacy as technological sources of knowledge and make it accessible to any Arctic citizen as part of an Arctic Educational Citizenship training proposed by this book.

Based on Figure 10.1 and on the five codes explained previously in Section 5.2, it is possible to visualize the in-depth levels of what it means 'to understand' and the main differences of conceptualization involved in this objective. It was also possible to have an overview of how the components of SD are dealt with and what is the adherence and role of energy (energy literacy or illiteracy) in the SD debate per case study. It was also relevant to realize characteristics related to the final purpose of these objectives and the levels of knowledge application in the four studied locations. The results suggest a significant lack of energy literacy and distinct educational approaches to SD and to energy education with no agreed cultural understanding in relation to these matters, certainly because development is influenced by economic models based on local energy models.

The interviews also revealed the main dissimilarities within the course objectives. Some of them are related to the indigenous and non-indigenous style of education and the respective methods and pedagogies practiced at the four studied locations (see Table 10.1). It is clear that a culturally inclusive curriculum is not the mainstream.

Based on the findings presented in Chapter 5 and through a comparative appraisal of common and different patterns it is possible to state that at Juniper University (JU) the course is more formal and theoretical, it follows the standardized definitions and principles shown in a formal theoretical framework largely adopted in Western academy. No concept or teaching of energy

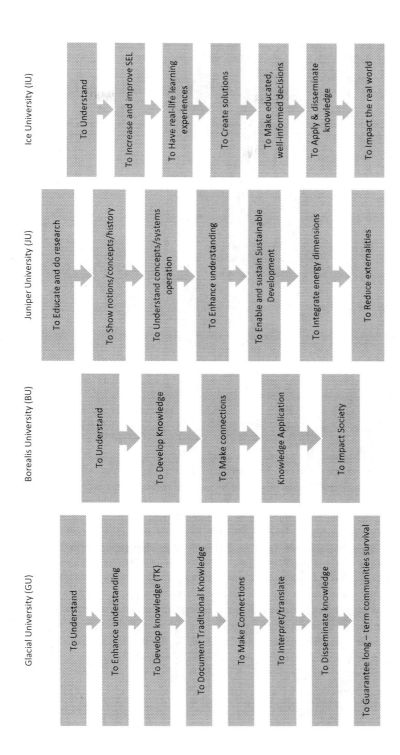

Figure 10.1 Comparative overview on charts summarizing the main course objectives in the four case studies

Table 10.1 Features of education in the case study universities

University	Features of Education
Juniper Univ. (JU)	Theoretical Formal Knowledge (no concept, no use of energy literacy) not a known notion
Borealis Univ. (BU)	Theoretical and Developmental Knowledge (energy literacy (not official term) but the notion is a component of ESD)
Glacial Univ. (GU)	Developmental, Practical Applied Knowledge (energy as an implicit element of Conservation of Biodiversity)
Ice Univ. (IU)	Developmental, Practical Applied Knowledge (sustainable energy literacy as an autonomous course as a realization of ESD)

literacy (as the main component of SDL) was identified and it was identified neither as a notion nor as a practice.

At Borealis University (BU) the main features were related to theoretical and developmental knowledge and energy literacy, despite not being an official term, is fully understood as a notion and it is a clear component of ESD. The emphasis to energy is connected to climate change and ecology more than social geography.

At Glacial University (GU), education shows a special emphasis on development, considered as an adaptive progression of the group or community. It also emphasizes the creation of practical and applicable knowledge with direct impact at individual and community level. SD and energy (an implicit component of SD relates to specific local projects) are studied through conservation of biodiversity performing a holistic, interdisciplinary, multicultural, and practical style of learning and education based on practical application and on community well-being. The focus is on the indigenous group and on its continuation as well as the local level, but with a concern about the global level participation and contribution envisioning the ideals in the concept of global citizenship. However, there is work to be done and more bridges to be built up between scientific knowledge and TK.

At Ice University (IU) the course is developmental, practical, with applied knowledge and skills to the everyday needs of individuals and communities in the real Arctic scenario. Sustainable energy literacy is an autonomous course working as a realization of ESD and searching for practical solutions to real problems in the field. It is a trend revealing a more democratic approach to educate the 'Arctic citizen' and it is in continuous development in the education set and, in the field, according to the dynamic characteristic of knowledge, oriented to the solution of complex and variable scenarios as the Arctic requires.

10.3.1 Levels of energy literacy

In terms of the structure of the energy courses offered in the selected universities and their levels of energy literacy, there are few studies discussing the topic in the Arctic. Additionally, in the literature there is no HE pedagogical model to assess levels of energy literacy. As presented in Chapter 7 (see Section 7.2) any model to be proposed depends primarily on the fundamental elements of DeWaters and Powers (2011, p. 1700) concept of an energy literate person, on the emancipatory capabilities approach from Sen (1992, p. 48), Nussbaum (2011, p. 33) and Freire (2000, p. 50), more specialized capabilities or 'geocapabilities' from Walkington et al. (2018, p. 11) to reach a representative model based on cognitive, affective, capability and behavioural subscales that integrates a typology based on knowledge, skills, attitudes and values, critical thinking, transformative learning, cultural capability, individual and collective responsibility, technology, intergenerational dimension, research, real-world practical application, and global-local contextualization: all essential components to tackle the Arctic adversities.

Based on the components inferred from the Arctic Pedagogical Model of Energy Literacy presented in Table 7.2 (Section 7.2), it was possible to create an inventory of the different components in the four university courses. Figure 7.1 shows comparative results based on cognitive, capabilities and behavioural course components that demonstrates high levels of energy literacy in relation to these three components at Ice University (IU). Here, energy literacy is embedded in undergraduate education aiming at discussing societal problems and technical solutions related to the unsustainability of current energy practices as well as the components of a sustainable energy model to be applied in practice.

In terms of capabilities, the course offered at Ice University (IU) tends to show the highest level of inventory components due to the practical orientation of the course at personal and community level, as students can apply knowledge and skills to their own real-life issues and on independent projects envisaging community well-being, autonomy, and empowerment. It also means an advanced level of interaction between science and TK (McCann, 2013).

Glacial University (GU) and Borealis University (BU) also record multiple inventory components as a proxy for capabilities in the course structure, whereas at Juniper University (JU) the emphasis on capabilities does not seem an expressive characteristic due to a more theoretical format. Capabilities denote the power for change through the individual.

In terms of cognitive characteristics Glacial University (GU) and Ice University (IU) shows relevant results in terms of the presence of cognitive components in their courses, followed by Borealis University (BU) and Juniper University (JU), demonstrating the relevance and the adherence of energy as a subject in the four case studies, but at different levels in practice.

In terms of behavioural features, including values, real life application and decision-making, Ice University (IU) has a more expressive level than Glacial

University (GU) and Borealis University (BU), notably due to the presence of a specific energy-literacy program and due to the focus on a behaviour change in terms of application of knowledge and skills to create solutions to current issues (i.e. smart grids, mini-grids, wind and solar community farms). It is important to remark that Juniper University (JU) demonstrated the least emphasis given to this component in comparison to the other case studies.

The Arctic Energy Literacy Comparative Assessment revealed that cognitive components and capabilities are being addressed mainly by IU, GU, and BU but behavioural components or the knowledge application require additional mechanisms of enhancement and assessment. This research has shown that the nature of the energy curriculum within sustainability/human geography is heterogeneous in the Arctic. In some courses, energy is more important than in others because there is still a fragmented view stimulated by the predominant current energy pattern based on fossil fuels. Energy literacy is an incipient field of studies in the Arctic and it is not well comprehended in some of the case study universities. The capability inventory seems useful as a means of comparison among the four case study universities in relation to the relevance energy has in the SD debate as component of an environmental impact or as a component of a sustainable energy model. These are two different positions, with two different dimensions, that reflect different levels of sustainability education less or more aligned to a sustainable energy model. These two different positions reflect two different energy realities (fossil and renewable energy). In this sense, this totally new inventory, that is a relevant contribution of this book and is called here, the Arctic Pedagogical Model of Energy Literacy is useful to promote a reflection and curriculum enhancement based on the transition from an environmental-only perspective (management of environmental impacts) to a sustainability perspective (durable wellbeing based on renewable energy for an alternative future), oriented to align course content, approaches and practices, on the side of staff, and to reflect on learning progress and capabilities acquired, on the side of learners. The application of this new model has the potential to be developed at program and module level by realigning the curriculum focus and practical application. It could also be accompanied by a reformed system of student evaluation more centred in the learners' acquisition and application, 'in situ', of geo-capabilities.

10.4 The structure of the energy curriculum from the four case study universities in the Arctic

10.4.1 Common factors

The most influencing factors that define the structure of the energy curriculum identified from the findings in the case studies refer to research-oriented courses, interdisciplinary characteristics, and resources/environmental management as part of the main course structure.

10.4.1.1 Research

The courses in the four universities targeted by the study were structured based on research.

Research and learning, depending on the specific context, are focused on themes associated to natural resources, climate change, cultural well-being, society, environment, biodiversity, land use, and energy production/use/efficiency. Research is considered an important skill to help students to collect data and use their data in their studies and applications.

10.4.1.2 Interdisciplinary characteristic

At Glacial University (GU) and Ice University (IU) interdisciplinary characteristic is an active component of the learning as the courses are considered more practical and explore the interrelationships and connections among factors in a more active way. Connections between TK and science are also materialized into specific databases and multicultural learning tending to trans-diciplinarity (Barnhardt & Kawagley 2005; Ilutsik, 2000; McCann, 2013; Vars 2007; UNESCO 2014).

At Borealis University (BU) and Juniper University (JU) interdisciplinary feature refers to the identified importance of making connections among factors even if it is not practiced. An interdisciplinary approach among subjects also happens considering that one course can combine modules so a student can study energy combining different intersections, like science, social environment, and the physical environment. Inter-disciplinarity characteristic would be strengthened with practice corroborating arguments by Robertson (2014).

10.4.1.3 Resources/environmental management

The courses present an intrinsic concern on how to manage natural resources and the environment regardless of the type of resources in question (land, energy, oil and gas, renewable energy). The management of impacts is a common concern in the courses analysed in the studied locations. At Glacial University (GU) and Ice University (IU) the emphasis of resources management is more oriented to the balance of communities' livelihoods and to enhance communities' capacity to perform sustainable ways of operating and adapting to change.

10.4.2 Distinctive factors

At Glacial University (GU) the course is structured based on 'non-traditional' components (meaning components not aligned to the Western course model) and a combination of place-based learning and distance learning. The course is structured based on experiences of lecturers and students as co-creators of the curriculum with features of trans-diciplinarity, multilinguism, and multiculturalism. Themes are studied under the viewpoint of TK with a special focus on documenting TK tending to curriculum integration by dissolving the boundaries between the conventional disciplines and organizing teaching and learning around the construction of meaning in the context of real-world problems (UNESCO 2014, p. 46).

A participatory approach within the cultural context, socialization, and knowledge dissemination within communities all seem to be essential structural elements maximized by information technology, making human encounters and dialogue a more meaningful experience. There is great concern on the thinking about values, culture, awareness, environment, change and adaptability, science, and culture. Teaching and learning are a cultural experience. The curriculum, pedagogies, and approaches are more interactive and holistic, but they are also predominantly reflective, participative, democratic, and inclusive. The type of curricula informs how TK can be used and be helpful for sustaining indigenous societies.

At Borealis University (BU) the most important characteristic of the curriculum is to make students understand the importance of participation to the study of geography and SD. A more participative approach and motivating participative pedagogies are important features of the course structure in order to stimulate critical thinking. It was emphasised that participation in class or under the form of societal impacts was mentioned but no specific example was provided apart from participation in national and international funded projects. This is justified by the employability dilemma. Energy is highly related to or studied within climate change; it does not have an autonomous branch of energy literacy. Reflection and application of concepts into practice are practices much desired at Borealis University (BU).

At Juniper University (JU) the curriculum focuses on the intersection of health and environment. The structure of the course shows energy as an implicit component because local economy is based on energy production and use. The curriculum has a specific focus on the intersections of social and physical environments, in other words, 'where people live and their health' and 'environmental management'. It is a lecture-based course covering topics related to economics, agricultural revolution, environmental issues, health, and quality of life, inclusive energy. Climate change is a sensitive topic causing tension, as there is some scepticism particularly amongst individuals from the energy industry. The approach has a systemic element – systemic thinking is a component of the course structure or the course approach and it is reflected in a discourse about the economic system and the socio-environmental impacts of the current economic system making a distinctive division into two 'invisible territories' or subjects: one is energy and the other is sustainability. One side focuses on the energy industry, production, and demand, but the other focuses on socio-environmental impacts and management. There seem to be a line dividing these two subjects and they do not mix. Despite the recognition of the importance of ESD and energy, this division is the cause of an implicit tension that is not addressed. The focuses of these two areas are different and it has negative implications especially under an indigenous cultural viewpoint.

The structure of the course at Ice University (IU) comprises core subjects and electives that allow students to specialize in the areas of their interest, but the course structure is focused on sustainable energy and energy literacy. The central object and approach of the course is to promote basic energy literacy to the highest possible number of people. The second most important objective is to promote real-world changes, real-world solutions, by the application of energy knowledge. The content is focused on the basic energy literacy, energy efficiency, renewable energy, and energy science with a component of socio-environmental impacts as the realization of Sustainable Development. Energy is an important pillar of ESD. Technology, contextualization, and practical life application represent a stimulus to student engagement in energy literacy and learning. The course is oriented to developed capabilities in students that become equipped to apply their knowledge into their own lives and on community level.

As presented in Chapter 1 (Section 1.1), energy continues to be a central factor in the historic human trajectory with significant implications to understand the SD process. Education has a role in promoting the transition from an environmental to a sustainability perspective, and important components like research, participative approach, interconnections among factors, multi-disciplinarity were identified in the course structures. However, the heterogeneity is a recurrent characteristic making it difficult to identify common patterns among the case studies beyond the ones identified. The structure of the energy curriculum is diverse and with different foci, levels of adherence and application considering the four case studies. This evidence reflects the local significance of energy, the predominant type of energy resource developed locally and how the society relates to energy. Ice University (IU) presents a specific energy literacy program with focus on making any person (indigenous or non-indigenous) knowledgeable in energy and oriented to forms of integrating energy (smart grids) in the Arctic and concerned with multicultural interactions and well-being. This is a relevant finding to promote further reflection about the consequences of not addressing energy with the same emphasis or not promoting further debate about how to use the local capacities towards a sustainable energy system.

10.5 The processes through which these curricula are designed and reviewed

It has been observed that curriculum design and change can be a complex process involving several kinds of influences. Based on the findings of this research indigenous curriculum design and review tend to be a co-reflective participative process, because it is about knowing how to live together (Delors 1996, p. 37) and sharing knowledge along the learning process in the field. This is, in fact, an important component of ESD. Lecturers and students are co-creators of the curriculum trying to consider more than one cultural perspective (indigenous and non-indigenous) and local-global realities in the course plans. Cultural activities shared along the learning journey and points of tension also contribute to improve and develop the curriculum that tends to be more creative, holistic, and adaptive in structure. This trend reflects the content, literature, approaches, and methodology adopted in the course emphasizing inquiry and contextualization.

The curriculum design and review when the considered environment is not exclusively indigenous tended to follow the local characteristics and background. Apart from an existing aspiration to make the process participative the curriculum is fundamentally designed and reviewed by the lecturer that tries to address the students' concerns, but it can happen in different moments at different levels of expectation, or in places simply not happen at all. This system of design and review is more aligned to a Western style and can create significant tensions when this curriculum is applied to mixed groups of learners that can feel without a proper voice to express their opinions, or without the right to be who they really are (Nussbaum 2011, p. 33) or feel that the curriculum is not adapted to indigenous ways of learning (pace of learning, amount of content conveyed, communication styles). However, the consensus corroborated by the students' interviews coding was of a more proactive curriculum oriented to practice (SD as practice).

The ways Arctic HE curricula are designed and reviewed are also heterogeneous revealing characteristics of the economic and cultural local realities varying the levels in which these curricula align to a sustainable model (of energy and of living). This is a point highlighted in the literature by Freire (2000, p. 50) when arguing that education and curriculum tend to follow the cultural settings of values and behaviours practiced by a particular society.

This book contributes to a reflection of how the curriculum design and review take place in the four case study universities and the importance of reflecting on how to promote a more inclusive, participative, and SD-aligned curricula to address the tensions of the 21st-century Arctic in the educational institutions and in society. It also calls attention to the relevance of a more creative, holistic, multicultural, practice-oriented, and integrated style of co-created and co-managed curriculum with a higher adherence to SD through which learners can feel comfortable about who they are and understand and help shaping alternative and collective sustainable futures based on geo-capabilities and responsibility to empower themselves as individuals and as a group. This process of reflection and curriculum change is only possible if local structures are opened to 'widen concepts of knowledge' instead of reproducing the same patterns and discourses of the existing power structures.

10.6 The structure of the energy courses offered in the selected universities and their adherence to ESD

10.6.1 Adherence to ESD

Authors who have focused their studies on the relationship between education and SD have felt the need to create typologies of ESD being delivered through HE curricula like Vare and Scott (2007, p. 3,4) framework of ESD1 and ESD 2, Scott and Gough (2003, pp. 113, 116) Type 1,2,3, Delors (1996, p. 37) pillars of education. In Chapter 2 and in Chapter 3 current models of ESD were presented and explained according to predominant ways of thinking and different levels of capacity of managing change and making decisions in relation to future scenarios.

Based on previous established research (Section 2.7) and adapting their lessons to an Arctic Higher Education based on its range of specific demands and characteristics, a new model of 'five types of ESD' was devised in this book (Table 7.5) by incorporating active components from St. Clair (2003), Mezirow (2009), Hein (1991), Kolb (1984), Henderson and Ikeda (2004), Haigh (2014, p. 14), Delors (1996, p. 37) and Walkington et al. (2018, p. 11) based on practical, experiential, and effective activities and skills for ESD realization. These levels assess adherence from a more passive position, in which SD is discussed on a more theoretical style passing by an intermediary level in which active knowledge and application is desired (but not yet being experienced) and would contribute to inform choices and decisions to Type 3, in which learning is adaptive and has the purpose of causing a tangible impact on how to live in the future. The real knowledge in action would be experienced when 'Education for Arctic Citizenship' is realized because when a learner is an Arctic citizen, he/she works for the common good, he/she creates and co-manages knowledge with a higher purpose than a job after the course. His/her knowledge will be directly and immediately applied, considering different perspectives of the local and regional stakeholders, and change is a reality under economic, social, environmental, and cultural pillars.

According to a new proposal which synthesizes and extends previous classifications shown in Table 7.6, *Type 1 (learning about SD)* starts from the premise that expert and learner are at different levels. Learning about SD denotes passivity by the learner and lecturer as the learning experience takes place on a more abstract and theoretical manner. An expert will tell the learner what he must learn and what he/she is. The real-world problems are fundamentally reduced to environmental issues and the hope is that technology applied to environmental management solves them all.

Type 2 (learning for SD) starts from the premise that the world has fundamental problems that are of different natures (i.e. social, political, environmental, economic, etc.) and these problems may be understood from a socio-scientific viewpoint and, possibly, by a more open-minded perspective that may include other forms of knowledge (i.e. TK) that will inform choices and decision-making among alternative futures. Learning for SD consists in willing and working for change focused on social pillar, consequently, application is desired, meaning that results (change) are desired but no one can assure that they will be achieved.

Type 3 (learning as SD) implies adaptive learning. Expert and learner work together for a tangible outcome (impact) by assuming that what represent knowledge is not adequate as 'end-states' cannot be specified and learning must be open-ended. These characteristics demonstrate the adaptive nature of Type 3 as knowledge will be generated to face the uncertainties and complexities inherent to life leading to reflective social learning and a reflection about the future. However, it is not clear how to measure the impact and tangibility of the outcomes.

Type 4 in this book called 'Education for Arctic Citizenship'. Through Education for Arctic Citizenship, multicultural experts and learners co-create knowledge and co-manage the learning experience for a common outcome as a collective exercise (socially and culturally inclusive) that is documented and passible of reflection. The assumption is that the concept of knowledge can be widen (broaden); knowledge is not static, mechanic or under pedagogic imperialisms, it considers more than one perspective, it is in constant adaptation as a dynamic system that can create new realities in different spaces and times. SD is in everything, it is the central concern in relation to the understanding of bio-geo-physical systems, food chains, socioeconomic systems, climatic systems, energy systems, natural resources management and it is embedded in the learning experience consisting in action in a SD context. SD is lived actively by expert and learner according to real-life challenges and complexities and it aims to inform everyday-life decisions to resolve real-world problems towards a state of well-being, freedom, and future existence.

Knowledge is a tool of empowerment for communities and societies to understand how to transform the reality for the Arctic citizens indistinctively of ethnicity by applying geo-capabilities. It represents more than learning to know and learning to do but learning how to live together in the learning site, in society, in the region by minimizing economic, social, environmental, and cultural tensions and by understanding the interdependence of living sustainably and responsibly. Responsibility in Arctic citizenship is a premise for Arctic sustainability. It means that the Arctic citizen uses his/her eco-literacy, energy literacy, and TK literacy to make educated decisions, to disseminate good practices and to interact constructively across cultural boundaries.

Type 5 of ESD is called *'Education for Global Citizenship'*, meaning that the actors are identified with the whole of humanity and use their skills, knowledge, and values to develop ways of living sustainably, responsibility, and peacefully. The global citizen demonstrates understanding, emotional intelligence, fairness, SDL or eco-literacy (energy literacy), geo-capabilities, and a refined sense of interconnectedness that leads to a creative co-existence of nature and humanity (Haigh 2014; Henderson & Ikeda 2004).

In this sense, Type 4 and Type 5 align or intersect with UNESCO's five pillars of learning (or five pillars of ESD) proposed by Jacques Delors (1996) in his book entitled 'The Treasure Within' published by UNESCO. The five pillars proposed by Delors (1996, p. 37) and explained in Section 2.7 aim at addressing tensions generated by social change like the universal versus the individual, the global versus the local, tradition versus modernity, long-term versus short-term considerations, competition versus equality, spiritual versus material, and the tension between the existing curriculum and innovative knowledge (Delors 1996, p. 37). Skills like comparing evidence, listening to different perspectives, understanding connections, and making judgements are essential for young people to make informed choices, reach consensus, to develop their full capacities, to actively participate in development and collaborate with others to make lifestyles more sustainable (Delors 1996; UNESCO 1990).

By interpreting the case studies, it is possible to conclude that type 1 is more expressively represented by Juniper University (JU) denoting a more formal Western-style of education in which learner is told what he is and what he must do. It is a more passive style of knowledge transmission from expert to learner based on impacts that must be managed and controlled according to economic, political, environmental, social, and managerial theories and principles.

Type 2 shows that courses from Borealis University (BU) and from Ice University (IU) are oriented to provide learning for SD, and application is desired; but the course from Glacial University (GU) emphasises that learning is a tool to empower people and facilitate decision-making towards the future (in the long run). However, Type 3 is clear to indicate that the courses from Glacial University (GU) and Ice University (IU) are operating on a more dynamic basis towards adaptation, complexity, unpredictability, application, and transformation denoting higher levels of adherence to ESD.

The outcomes obtained when assessing the courses adherence to ESD by using the 'Five types ESD model' resulted in the realization that Glacial University (GU) and Ice University (IU) courses are substantially more aligned to an active and effectively applied knowledge in the field by showing more refined levels of adherence to ESD according to the Education for Arctic Citizenship components and, consequently associated to ESD Type 4, where the emphasis is life-long learning leading to real-life learning experience and tangible outcomes within a multicultural environment.

However, there are still elements to progress and tensions to address. At Glacial University (GU), the focus of learning is on the indigenous group. Learning is co-produced and co-managed to promote adaptation well-being and intergenerational survival. Knowledge represents empowerment to overcome ethical tensions, to produce social (local) impacts and transformation for a durable model of existence. Knowledge production is a cultural process that is expanding from a local to a global level through the understanding that local changes represent global changes. SDL is a more integrated and comprehensive process as energy is part of a biological diversity research and analysis. In some interviews, elements of Type 5 were identified as some interviewees demonstrated a sense of compassion and global conscience beyond the cultural group.

At Ice University (IU) the main characteristic that defines Type 4 ESD adherence with elements of Type 5 ESD is the real-life learning experience as knowledge is immediately applied in the every-day life complex problems and decision-making to create a positive and efficient impact on well-being and sustainability. SDL (eco-literacy) and energy literacy are oriented to everyone in any occupation or position as all people are responsible and should know how to make well-educated decisions on how to use renewable resources to create efficient solutions to real-life complex problems based on a durable model of existence. The curriculum is not completely adapted to a multicultural environment as it does not contemplate TK, but different groups (indigenous and non-indigenous) experience the same real contexts and have to join efforts to address them using science and tradition in practice. The knowledge is applied in real-world projects to benefit local and regional communities.

The ESD Adherence Comparative Assessment, shown in Figure 7.2, is based on a new model (ESD Adherence Parameters) inferred from the primary and secondary data and an important contribution of this book showing trends of Arctic education. It confirmed the heterogeneity of characteristics of the courses under study as to ESD alignment in the Arctic. The assessment reflects exactly the features of the local societies and their orientations in relation to economic, environmental, social, and cultural balances.

In the location where the economic aspect is more prevalent the trend in education is of Type 1 (learning about SD) because sustainability consists in managing environmental hazards and impacts through technology and energy production (regardless of the source) is seen as a priority to maintain the economic system. In general, there are two groups of stakeholders: the sustainability people and the energy people, they do not mix as they present contrasting interests and values, consequently contrasting behaviours. Type 2 or learning for SD denotes that there is more than only the economic pillar. These educational institutions tend to consider social and environmental concerns as part of achieving a more balanced position based on an open mind-set when assessing choices and making decisions. Learning for SD means that the institutions envisage a balance and they pursue SD as a goal. The educational institutions showing Type 3 (learning as SD) are those that understand the adaptive nature of knowledge and the power of knowledge in promoting local social adaptation to complexities. There is a collaborative spirit of development to achieve tangible outcomes and envisage alternative futures. These universities have significant cultural predominance and a specific concern regarding unsustainable practices. Type 4 or Education for Arctic Citizenship represents the crucial innovative ESD parameter of this book by incorporating the multidisciplinary and multicultural foundations for a long-term durable model of Arctic wellbeing. The model is useful to promote a reflection and curriculum assessment and enhancement based on the notion of citizenship that realizes the five pillars of ESD proposed by Delors (1996, p. 37) meaning that the courses under Type 4 promote learning to know, to do, to live together, to be and to transform oneself and society by creating multicultural educational spaces. The application of Arctic Citizenship Education Model can re-orient several contemporary discussions, globally and locally, in academy and in organizations as how to make societies pro-active in addressing climate change, implement new sustainable energy systems and wellbeing.

10.7 Multidisciplinary and multicultural perspectives of a sustainable energy system in Arctic Higher Education curricula

Energy availability and access has a significant influence on how societies operate and interrelate leading to the idea that energy is a geographical concept as its production and use reveals different complex dynamics in distinct geographical contexts. These contexts have determined the types of resources explored, the volumes extracted, and the extension of side effects generated by these explorations. These particular dynamics have impacts on the environment, on the economics, on local communities but also on the ways these elements interrelate and are conveyed in HE. This interdependence of multidisciplinary elements raises an extraordinary net of complexity to individuals and societies to understand and to operate under natural limits within pre-stablished economic and cultural systems.

The Arctic context, in its profound complexity and sensitiveness, provides a very special stage of real-world interactions and extreme experiences to be understood and to be dealt with in everyday life, in all senses. Extreme experiences related to the local economies, local environment and local societal issues, just to juxtapose this scenario to the TBL, demonstrated the need of multicultural competences that are not being considered in the way the SD concept is currently applied. This is an analogy representative of a current TBL static model currently limited to these three pillars that data, in this book, proved to be insufficient as the real-world interactions are highly more dynamic and deeply more complex, more interdependent, intricate and difficult to understand and disseminate as knowledge than the compartmented and rhetoric style can provide. This context happens this way because of the interdependent layers of complexity among dynamic interactions cannot be perceived by the human being, or because these interactions don't receive the necessary attention by stakeholders in their daily consumerist attitudes or, even, because of the intertemporal feature of these interactions that may not be apprehensible by the human mind in its temporal and limited existence.

The multicultural component of this Arctic dynamic is evident and necessary, as people, regardless of their origin and social status, will need to adapt to the rapid transition that is in course. It means that this geographical and physical change will need to be accompanied by a change in the mind-set, behaviour, and quality (inclusiveness) of decision-making. The rhetoric whose discourse is fundamentally based on environmental impacts and how technology can save the planet has proved not to work in every geographical area of the planet. The complexity of real-world problems of different natures require a redefinition of society's relationship to the environment (as a home or as an exploration site); to energy (production, use and literacy of sustainable or unsustainable models); to economy (how to produce and use according to the Earth's capacity systems); to culture (how to address different worldviews on SD and sustainable energy system); mainly in highly sensitive areas like the Arctic. This redefinition requires a consistent and systematic review of HE curricula to promote a better integration of energy literacy components into undergraduate programs as a central element of SDL, the development of geo-capabilities under a model of Education for Arctic Citizenship as the social Arctic environment is naturally multidisciplinary and multicultural in its complexity and richness.

The fact that energy is hidden or headlined as ecology, environmental impacts, environmental management, or even, as climate change in several courses, represents a strong justification to make the case for energy being studied as a central theme of SD, and to promote efficient sustainable energy literacy for all, because it is not possible to tackle large scale problems without a concerted action from well-informed individuals and communities about the causes of the current Arctic changes and the unsustainable models in practice.

The intent of this book is to promote means to understand the processes of learning (even where the learning pathways and pedagogic approaches are different) in an era of transformation mainly caused by energy exploration/use and to extend understanding of the causes of this large-scale change instead of tackling the consequences or impacts only via simplistic and narrow curricula. It is a time to be proactive in addressing the causes, the social tensions and the contradictory discourses and interests by rethinking the processes of learning and reassessing the deficiencies that occur within and at the intersection of diverse world views and knowledge systems through well-defined parameters aligned to a sustainable model of existence. This study intended to provide a comparative analysis of experiences derived from four Arctic educational contexts. The case studies represent a sample of the Arctic Higher Education linked to specific local socioeconomic-environmental-cultural and energy system structures.

The findings of this research indicated a lack of sufficient models (from the literature) to assess the levels of adherence of courses in relation to ESD and energy literacy. This book argues the necessity to re-align Arctic Higher Education to specific competencies and intelligences necessary to proceed to a large-scale adaptation process that involves indigenous and non-indigenous peoples in the same Arctic scenario. This process would benefit if cognitive, affective, behavioural skills, and knowledge would be connected to a range of specific capabilities privileging critical thinking, cultural competence for SD, complexity-thinking capabilities towards behavioural change and real-life practical knowledge application. This is the reason why the Arctic Pedagogical Model of Energy Literacy was designed, derived from the data collected in this research. This is a model with broad applicability in other Arctic nations, but also in non-Arctic countries in need of inclusive and decolonized curricula to provide access for indigenous students in Higher Education.

In terms of ESD, the findings demonstrated that the knowledge, skills, and competences required to live and operate in the Arctic require a more comprehensive range of competences or intelligences aligned to a superior level of critical reflection oriented to long-term (intergenerational) collective outcomes based on multicultural respect and responsibility. This is the genesis of the Arctic Citizenship Educational Model that goes beyond participation patterns of education to set the learners in the real-life geographical contexts and learning sets to resolve daily problems with impacts on daily lives.

Results were presented following the ESD Adherence Comparative Assessment Graph in Chapter 8, demonstrating that there are courses aligned to the Education for Arctic Citizenship model where Arctic knowledge is being produced by the encounter of science and TK, and where multicultural experts are co-creating and co-managing active, open, practical, systemic, adaptive knowledge oriented to 'functionings', and to intergenerational outcomes. This is the first time such kind of evaluation has been done and one area for future research may be to monitor its application to verify if this adherence to ESD is being enhanced, in these four case studies and, perhaps, apply the model to assess other Arctic educational institutions. By focusing on the levels of adherence to ESD and to energy literacy, it is possible to enhance the chances of success in tacking the Arctic issues of climate change, adaptive process and multicultural cooperation and co-management. The application of these models presented in this book can also be part of global educational strategies to evaluate the courses alignment to ESD and SEL in other bioregions as a pro-active measure of conducting civilization to a more sustainable path of existence in the long run.

10.8 Limitations of the study and contributions for further research

This section outlines the limitations and contributions of the study. It is important to point out that to embed ESD in educational institutions it is necessary to first identify the levels of adherence of curricula to the main ESD components and it is also important to investigate the characteristics of the energy curriculum and how it is delivered to students. Energy is the central component of SD with important geographical implications to economies, societies, and natural environment. Consequently, energy is a key component of ESD, despite the narrow understanding of this fact and the limited presence of energy in sustainability/human geography-related programmes.

This research was conducted in a very particular context, the Arctic. A place where limitations for human beings are the rule, due to harsh weather and temperature conditions, remoteness, off-grid communities, limited transport systems and limited hours of light. However, the researcher has been working in this environment for more than 18 years and developed a systematic knowledge of how to operate in unusual conditions not being these elements the main research limitation. The main limitation experienced by the researcher in this study was related to the difficulty of keeping the confidentiality of the case study institutions and participants as per their specific and, sometimes, unique features regarding practices, activities, and styles of education. Several important materials, bibliographies, documents, part of the data collection revealing practices and pedagogies could not be included in the study for confidentiality reasons. It was a careful and accurate approach demonstrated by the researcher who followed strictly the established ethical guidelines proposed by the ethical committee.

The research was also conducted in a mixed research population environment where non-indigenous people and indigenous people were interviewed by the researcher. As a relatively experienced researcher familiarized with the multicultural environment of the areas where the study was conducted, the researcher established her role during the research process as a facilitator, an open-minded listener, a trust-builder aware of the fact that cultural factors are particularly important, given the risks of ethnocentric predisposition and cultural imperialism (Wilson 1992). As the researcher had worked previously in other Arctic projects involving multicultural groups, she understands the importance of minimizing bias and polarized approaches (Bryman 2016, p. 216) in order to provide a fruitful space for cultural analysis (Fang & Faure 2011) and cultural voice in relation to multicultural tensions from the field. This way, the researcher played a role as a social actor in relation to the studied groups (Hill & McLennan 2016, p. 670; Kloos 1969) when promoting reflection about SD, energy literacy, and ESD at the point that several participants asked to be informed about the findings when the study was finished and published. Researcher bias is possible as researchers are social actors and it is impossible not to bring something to the data as an interested researcher.

Despite being a small sample, the case studies reflected extreme contextual and complex characteristics of the places and communities researched. Economic models based on specific energy systems influence HE education. Places developing different energy sources have different HE dynamics in relation to ESD and EL. All Arctic citizens should have the right to be part of SD in practice. The institutions were selected due to the most representative educational dynamics in relation to the topic under study to formulate valid counterpoints in relation to the concepts studied. It is accepted that there are other educational initiatives in different parts of the Arctic, but the aim was to present a diversity of perspectives from locations with different cultural and economic features (influenced by energy production/use). Another fundamental contribution and opportunity for further research resides in the fact that the models proposed in this book can be applied to evaluate other courses adherence to ESD, to energy-literacy levels and to understand the main competences, intelligences and outcomes that educators and students in different geographical areas want to emphasise and to align to. Based on this combination of elements new curricula, practices, and pedagogies can be re-aligned or re-designed representing future areas of development. The models can be applied in other areas of the Arctic like Greenland, or in different geographical areas of the globe that are pursuing a change in the educational paradigm towards integrating ESD.

10.9 Conclusion

The starting point in this book journey was the premise that the way energy is produced and used has an intrinsic relationship with societal development and the natural world. The critical element of change in the Arctic is energy production/use in sensitive areas impacting human and natural systems and the (in)capacity of understanding and addressing, the scale of the consequent transition.

People, regardless of their origin and ethnicity, will need to adapt to the range of changes and the energy transition in Arctic and non-Arctic nations. In the Arctic, the level of knowledge and skills required are higher than only economic, social, and environmental aspects of the TBL. The TBL pillars must be balanced and inclusive of all the Arctic inhabitants (indigenous and non-indigenous) as they need to have the right to understand the impacts of energy exploration on natural resources (and land) and they must have the right to decide on how to manage and take well-educated decisions about the type of energy resources to use and how to use them efficiently according to their history, traditions, practices and ways of life.

The current discourse in Arctic areas with past or present predominance of fossil fuel exploration is about technology, environmental impacts, and environmental management. These elements are insufficient to reflect the values of SD or of an Arctic Citizenship. Education is an extraordinarily powerful tool to help the Arctic peoples to understand and adapt to the changes of the current transition. This transition needs to be accompanied by a powerful and culturally inclusive curriculum to provide the necessary practical knowledge and experience adapted to different ways of learning and able to forge a future sustainable energy system that is fundamental for

Arctic and non-Arctic nations to shape their desired futures. Education oriented to develop an adaptive capacity for communities to respond to the scale of the current transition will be essential to secure peoples' continuation, culture, and livelihoods, it is more than education for employability. Energy literacy as the main part of SDL should be performed under transdisciplinary and multicultural perspectives to effectively promote a broader understanding of the scale of causes and consequences of the civilization's energy exploration activities in the last two centuries.

Education needs to go beyond the current patterns, approaches, pedagogies, assessments to include communities in this process of adaptability and in the current debate. It would be a real sign of Arctic Citizenship. However, it is also important to observe that adaptability also depends on other factors like adequate infrastructure and investments. Funding availability is also a factor affecting adaptive capacity. A more adaptive form of education combining TK and scientific knowledge can certainly reduce vulnerability and strengthen Arctic communities' resilience and support-capacity to implement stronger co-management adaptive strategies in response to transition as some of the student's and lecturers' reports from the data collection demonstrated are being done in the Arctic. The indigenous peoples are interested and opened to learn science, SD, energy, and climate change. They are opened to interact, exchange and participate in the current debate through research, documenting TK, and building education and curricula oriented to sustainable ways of development.

Development that is durable means the ability of living a life according to a balanced functioning, well-being and freedom according to Nussbaum (2000, p. 72), Sen (1999, p. 75), Freire (1998, p. 33) and Delors (1996, p. 37). It is not possible to exercise functionings, welfare, and freedom without truly understanding the SD concept from a range of different perspectives and be able to apply it. Education oriented to values, context-specific factors (geo-multicultural competences), and with a vision towards geo-capabilities contributes to citizens' capacity of making choices for sustainability.

The Arctic peoples have not been allowed to experience all the human capabilities or geo-capabilities related to making choices for sustainability, creativity to be productive in an economic and cultural context of autonomy. This book argues that a multi-culturally accommodating curriculum in the Arctic based on geo-capabilities and citizenship would benefit Arctic learners by recognizing multicultural assets, applying cross-cultural teaching practices, mediating tensions, and enriching exchanges based on what both perspectives can offer to learners. The curricular role of geography goes beyond the contextualization to provide how to think geographically and multi-culturally as physical and human aspects are part of the same spaces. Geographical studies and energy literacy are fundamental to promote connectivity between the natural and social worlds and to empower the young Arctic community. Curricula that engage indigenous and non-indigenous peoples and align to the proposition of an Arctic Citizenship is greatly demanded to assure that the Arctic citizens understand the different perspectives on SD, co-create paths for a sustainable model of co-management and co-existence.

The research outcomes revealed that a traditional Western education, emphasizing the economic pillar of the TBL, does not provide the necessary adaptability mechanisms to face the scale of the current Arctic and global transition. A culturally inclusive curriculum that includes contextual, multicultural, and multidisciplinary components with higher levels of adherence to ESD, energy literacy and geo-capabilities seems more adequate to invite different groups of people to engage into a motivating and meaningful dialogue, learning experience, and knowledge application that a sense of regional and global citizenship will require. The challenges presented by the rapid and complex transitioning period affecting all societies could be minimized if energy efficiency and renewable energy sources are given a chance to consolidate a more durable model of local development. This is an ongoing process.

Multicultural education and research have experienced a considerable growth in several parts of the Arctic recently and a multicultural accommodating curriculum is an important measure to enhance a meaningful HE access and engagement to First Nations, Sámis, Inuit, Metis and American indigenous peoples. An emerging group of indigenous scholars across the Arctic is contributing in this accommodation process with views of how knowledge is constructed and disseminated (Scollon 1981). The Arctic Citizenship Educational Model, presented in this book, will be of great value in this current co-building process.

TK plays a significant role in socio-environmental management because it is defined, in essence, by the same characteristics of what is considered sustainable. Documentation and transmission of traditional expertise represents a powerful tool for development and preservation contributing to interactions with the scientific knowledge. Non-indigenous nations benefit of learning TK because *birgejupmi* (UNPFII9 2010) *and isumatturuk* (Webster & Zibell 1970) (livelihood, survival capacity) can remind the Western civilization how to live with others and how to reflect on its survival capacity. Too much emphasis on economic profit would fragment the traditional notion of survival capacity. This concept needs a balance between the natural, cultural environment and society to meet people's physical, mental, and social health, to allow people to exercise functionings, to be who they are and to achieve their best potential in life according to their cultural asset. Development needs to be truly sustainable locally and globally and it requires competences, intelligences, and transposing perspectives.

Cross-cultural *praxis* requires experiential learning and collaboration in academic contexts in which individuals can learn to address the tensions of and enlarge a type of third space (Aoki 1991) where people know how to live together and can exercise the functionings and their freedom. A space where oral tradition can be understood pedagogically, and science can be understood in practice. This space needs to be a safe space for multicultural students to connect with others, being themselves, without the demands and tensions of the Western dominant culture or the employability dilemmas. The creation of multicultural educational spaces and curricula in HE is an ambitious proposition, that requires holism, authentic good will for mutual learning and civilizational well-being.

References

Aoki, TT 1991, 'Teaching as in-dwelling between two curricular worlds', in T Aoki (ed.), *Inspiring curriculum and pedagogy: talks to teachers*, University of Alberta, Edmonton, pp. 7–10.

Arruda, GM 2018, *Renewable energy for the Arctic: new perspectives*, Routledge, Abingdon.

Arruda, GM & Krutkowski, S 2017a, 'Arctic governance, indigenous knowledge, science and technology in times of climate change: self-realization, recognition, representativeness', *Journal of Enterprising Communities: People and Places in the Global Economy*, vol. 11, no. 4, pp. 514–528.

Arruda, GM & Krutkowski, S 2017b, 'Social impacts of climate change and resource development in the Arctic: implications for Arctic governance', *Journal of Enterprising Communities: People and Places in the Global Economy*, vol. 11, no. 2, pp. 277–288.

Ascher, M 2002, *Mathematics elsewhere: an exploration of ideas across cultures*, Princeton University Press, Princeton, NJ.

Atleo, MR & Fitznor, L 2010, 'Aboriginal educators discuss recognizing, reclaiming and revitalizing their multi-competences in heritage/English language use', *Canadian Journal of Native Education*, vol. 34, pp. 13–34.

Baker, S, Kansis, M, Richardson, D & Young, S 1997, *The politics of sustainable development*, Routledge, London.

Barnhardt, R & Kawagley, OA 1999, 'Education indigenous to place: western science meets indigenous reality', in G Smith & D Williams (eds.), *Ecological education in action*, State University of New York Press, New York, pp. 117–140.

Barnhardt, R & Kawagley, OA 2005, 'Indigenous knowledge systems and alaska native ways of knowing', *Anthropology & Education Quarterly*, vol. 36, no. 1, Indigenous Epistemologies and Education: Self-Determination, Anthropology, and Human Rights (March, 2005), pp. 8–23.

Battiste, M & Hernderson, J 2000, *Sa'ke'j Youngblood. Protecting indigenous knowledge and heritage*, Purich Publishing, Saskatoon, SK.

Berkes, F 2008, *Sacred Ecology*, 2nd edn, Routledge, New York, p. 3.

Bitz, C, Blockley, E, Kauler, F, Petty, A, Massonet, F, Arruda, GM, Sun, N & Druckenmiller, M 2016, *Post-season report publishing*, Washington, DC, viewed January 2017, www.arcus.org/sipn/sea-ice-outlook/2016/post-season.

Boulton, J, Bowman, C & Allen, P 2015, *Embracing complexity: strategic perspectives for an age of turbulence*, Oxford University Press, Oxford, UK, pp. 1, 10, 11, 20, 29, 35, 63, 66, 117.

Bryman, A 2016, *Social research methods*, 5th edn, Oxford University Press, Oxford, pp. 3, 78, 216.

Cortese, A 2003, 'The critical role of higher education in creating a sustainable future', *Planning for Higher Education*, vol. 31, no. 3, pp. 15–22.

Delors, J 1996, *Learning: the treasure within*, Report to UNESCO of the international commission on education for the twenty-first century, UNESCO, Paris, p. 37.

DeWaters, JE & Powers, SE 2011, 'Energy literacy of secondary students in New York State (USA): a measure of knowledge, affect, and behavior', *Energy Policy*, vol. 39, pp. 1699–1710.

Eglash, R 2002, 'Computation, complexity and coding in Native American knowledge systems', in JE Hankes & GR Fast (eds.), *Changing the faces of mathematics: perspectives on indigenous people of North America*, National Council of Teachers of Mathematics, Reston, VA, pp. 251–262.

Fang, T and Faure, GO 2011, 'Chinese Communication Characteristics: a Yin Yang Perspective.' *International Journal of Intercultural Relations*, vol. 35, no. 3, pp.320–333.

Fernando, JL 2003, 'The power of unsustainable development: what is to be done?', *The Annals of the American Academy*, vol. 590, pp. 6–31.

Freire, P 1998, *Pedagogy of freedom: ethics, democracy and civic courage*, Rowan & Littlefield, Lanham, MD, p. 33.

Freire, P 2000, *The pedagogy of the oppressed*, Alley Cat Editions, New York, NY, p. 50.

Haigh, M 2014, 'From internationalisation to education for global citizenship: a multi-layered history', *Higher Education Quarterly*, vol. 68, no. 1, pp. 6–27, 14, 16.

Hein, G 1991, Constructivist learning theory, viewed 20 June 2018, www.exploratorium.edu/ifi/resources/constructivistlearning.html.

Henderson, H & Ikeda, D 2004, *Planetary citizenship: your values, beliefs, and actions can shape a sustainable world*, Middleway Press, Santa Monica, CA.

Hill, CM & McLennan, MR 2016, 'The primatologist as social actor', *Etnográfica*, vol. 20, no. 3, pp. 668–671.

Hove, H 2004, 'Critiquing sustainable development: a meaningful way of mediating the development impasse?', *Undercurrent*, vol. I, no. 1, pp. 48–54, 53.

Ilutsik, E 2000, 'Traditional Yup'ik Learning', in S Stephens (ed.), *Handbook for culturally responsive science curriculum*, Alaska Science Consortium and the Alaska Rural Systemic Initiative, Fairbanks, p. 30.

Kloos, P 1969, 'Role conflicts in social fieldwork', *Current Anthropology*, vol. 10, no. 5, pp. 509–523.

Kolb, DA 1984, *Experiential learning: experience as the source of learning and development*, Prentice-Hall, Upper Saddle River, NJ.

Longman, M 2014, 'Aboriginography. A new decolonized aboriginal methodology', in G Guttorm & SR Somby (eds.), *Diedut, V.3*, Sami University College, Alta, pp. 16, 17.

Magni, G 2016, Indigenous knowledge and implications for the sustainable development agenda. Education for people and planet: creating sustainable futures for all. Background paper prepared for the 2016 Global Education Monitoring Report. UNESCO, p. 5, https://pdfs.semanticscholar.org/46cc/5a5297743f31abfacce0c8ff89754bdbfe18.pdf.

McCann, H 2013, Local and traditional knowledge stewardship: managing data and information from the Arctic. National Snow and Ice Data Center, viewed 19 November 2016, www.arcus.org/witness-the-arctic/2013/2/article/19956.

Meadows, M 2009, 'Walking the talk: reflections on indigenous media audience research methods', *Journal of Audience Studies*, vol. 6, no. 1, pp. 118–136.

Mezirow, J 2009, 'An overview of transformative learning', in K Illeris (ed.), *Contemporary theories of learning*, Routledge, Abingdon, pp. 90–105.

Nussbaum, M 2000, *Women and human development: the capabilities approach*, Harvard University Press, Cambridge, MA, pp. 70–77, 72, 73, 74.

Nussbaum, M 2011, *Creating capabilities: the human development approach*, Harvard University Press, Cambridge, MA, pp. 33–34, 33.

Orr, DW 1992, *Ecological literacy: education and the transition to a postmodern world*, SUNY Press, Albany, NY, p. 210.

Orvik, J & Barnhardt, R 1974, *Cultural influences in Alaska Native Education*, Center for Northern Education Research, University of Alaska Fairbanks, Fairbanks, AK.

Poppel, B 2015, *SLiCA: Arctic living conditions – living conditions and quality of life among Inuit, Sami and indigenous peoples of Chukotka and the Kola Peninsula*, Nordic Council of Ministers, Denmark, p. 67.

Robertson, M 2014, *Sustainability principles and practice*, 1st edn, Routledge, Oxford, p. 7.

Ryan, A 2011, ESD and holistic curriculum change, pp. 3, 5, viewed 12 December 2016, www.heacademy.ac.uk/resources/detail/sustainability/esd_ryan_holistic.

Scollon, R 1981, *Human knowledge and the institutions knowledge: communication patterns and retention in a Public University*, Center for Cross-Cultural Studies, University of Alaska Fairbanks, Alaska.

Scott, WAH & Gough, SR 2003, *Sustainable development and learning: framing the issues*, Routledge, Falmer and London, pp. 113, 116.

Sen, A 1992, *Inequality re-examined*, Claredon Press, Oxford, pp. 5, 39, 40, 48.

Sen, A 1993, 'Capability and well-being', in M Nussbaum & A Sen (eds.), *The quality of life*, Clarendon Press, Oxford, pp. 30–53, 31.

Sen, A 1999, *Development as freedom*, Oxford University Press, Oxford, pp. 19, 75, 87.

Sillitoe, P 1998, 'The development of indigenous knowledge: a new applied anthropology', *Current Anthropology*, vol. 39, no. 2, pp. 223–252.

Smith, LT 1999, *Decolonizing methodologies: research and indigenous peoples*, Zed Books, New York.

Srinivasan, R 2006, 'Indigenous, ethnic and cultural articulations of new media', *International Journal of Cultural Studies*, vol. 9, no. 4, pp. 497–518.

St. Clair, R 2003, 'Words for the world: creating critical environmental literacy for adults', *New Directions for Adult and Continuing Education*, vol. 2003, no. 99, pp. 69–78.

Stephens, S 2000, *Handbook for culturally responsive science curriculum*, Alaska Science Consortium and the Alaska Rural Systemic Initiative, Fairbanks, AK, p. 10.

Sterling, S 2003, 'Whole Systems Thinking as a basis for paradigm change in education: Explorations in the context of sustainability', Unpublished doctoral dissertation, University of Bath, Bath.

The Energy Efficiency Partnership 2013, Alaska energy efficiency, viewed 5 December 2017, www.Akenergyefficiency.org/.

Tuhiwai Smith, L 1999, *Decolonizing methodologies research and indigenous peoples*, Zed Books Ltd., London and New York.

Tuhiwai Smith, L 2005, 'On Tricky ground – researching the native in the age of uncertainty', in NK Denzin & YS Lincoln (eds.), *The Sage handbook of qualitative research*, 3rd edn, Sage Publications, Inc., London, pp. 85–108.

UNESCO 1990, *World declaration on education for all*, Secretariat of the International Consultative Forum on Education for All, New York.

UNESCO 2014, *Education strategy 2014–2021*, UNESCO, Paris, p. 46, viewed 2 December 2018, www.natcom.gov.jo/sites/default/files/231288e.pdf.

UNPFII9 (Ninth Session of the UN Permanent Forum on Indigenous Issues) 2010, *Report on the ninth session*, 19–30 April 2010, UN Headquarters, New York.

UNWCED 1987, Report of the world commission on environment and development, General Assembly Resolution 42/187, 11 December 1987, p. 10, viewed 30 March 2018, www.un.org/documents/ga/res/42/ares42-187.htm.

Vare, P & Scott, WAH 2007, 'Learning for a change: exploring the relationship between education and sustainable development', *Journal of Education for Sustainable Development*, vol. 1, no. 2, pp. 3, 4.

Vars, LS 2007, 'Hvorfor bør man og hvordan kan man bevare samenes tradisjonelle kunnskap?', in JT Solbakk (ed.), *Árbevirolašmáhttu ja dahkkivuoigatvuohta – Tradisjonell kunnskap og opphavsrett –Traditional knowledge and copyright*, Sámikopija, Karasjok, Sámi University College, pp. 123–166.

Walkington, H, Dyer, S, Solem, M, Haigh, M & Waddington, S 2018, 'A capabilities approach to higher education: geocapabilities and implications for geography curricula', *Journal of Geography in Higher Education*, vol. 42, no. 1, pp. 7–24.

Webster, DH & Zibell, W 1970, *Iñupiat Eskimo Dictionary*, Department of Education, University of Alaska, Alaska.

Wheatley, M 1999, *Leadership and the new science: discovering order in a chaotic world*, 2nd ed., Berret-Koehler Publishers, San Franscisco, CA, p. 10.

Wilson, K 1992, 'Thinking about the ethics of fieldwork', in S Devereux & J Hoddinott (eds.), *Fieldwork in developing countries*, Harvester Wheatsheaf, London, pp. 179–199.

Youngblood, M 1997, *Life at the edge of Chaos*, Perceval Publishing, Dallas, TX, pp. 10, 27.

Index

Note: Boldface page numbers refer to tables and italic page numbers refer to figures. Page numbers followed by 'n' refer to endnotes.

Aberley, D 98
aboriginography 49
ACSYS *see* Arctic Climate System Study (ACSYS)
adaptation 5, 7–8, 84, 94, 114, 206, 210, 215, 218, 225, 228, 247; for Arctic societies 38–41; curriculum 155; education and 41–4
Adger, WN 39
Agenda 2030 57, 211, 215
Agius, P 50
agricultural revolution 1–2
Alaska 34, 38; energy education 69; energy systems 32; as fossil fuel producer 33; hydropower resources 32, 37; indigenous peoples 44, 47; languages of 47; natives 46; petroleum resources 32
Alaskan Arctic 32, 37
Alaska Native Claims Settlement Act 46
Alberta Emerald Foundation 50
Allen, P 85
AMAP *see* Arctic Monitoring and Assessment Program (AMAP)
Arctic: basins, oil and gas 29; climate change 5, 36; communities 44; education as adaptation tool 41–4; education in multicultural societies 44–53, 45, **46**, 47; element of change in 66; energy and 6–8, 28, 29–31; energy production/use 249; energy transition and governance 31–3; ethnical diversity in 48; fossil fuel in 29, 31, 33; geo-capabilities 250; governance 34–5; holistic dimension of 44; indigenous peoples 35, 42, 43, 46; multicultural component of 246; relevance of UN SDGs to 210–14; renewable energy projects in 37, 38; sustainable energy literacy 67–9; transformation, political decision-making on 33–5
Arctic Centre for Sustainable Energy 69
Arctic Circle 84, 103
Arctic Citizenship 213, 214, 228, 232, 242, 249, 250
Arctic Citizenship Education Model 174, 228, 245, 247, 249, 251
Arctic Citizenship Model 174, **175**–7, 247
Arctic Climate System Study (ACSYS) 5
Arctic Council 38
Arctic Council Sustainable Development Working Group (SDWG) 43
Arctic Educational Citizenship 232
Arctic energy curriculum 108, 134; Borealis University 126–8; environment and natural resources 117–18; Glacial University 123–6; Ice University 130–2; intergenerational impacts 120–1; interpretivist method 108; Juniper University 128–30; key objectives of 109–11, **110**; lecturers' code names **109**, 108; lecturers' interviews 133–5; natural resources management 118–20; nature of 109–23; objectives of 133–5; students aware of SD 114–17; theoretical to practical learning mode *134*; understanding of SD concept 111–14
Arctic energy curriculum, structure of 137; Borealis University 142–3;

capabilities developed in students 147–52; curriculum characteristics **138**; curriculum design 137–8; Glacial University 140–1; Ice University 145–7; interdisciplinary structured courses 139; Juniper University 143–5; research as structural course component 138–9; resources/environmental management 139–40
Arctic Energy Literacy Comparative Assessment *170*, 236
Arctic environment 29, 38, 43, 246; climatic changes 111; complexity theory and 83–4
Arctic Higher Education 8, 41–2, 50, 115, 211, 221, 241, 246; curriculum 1, 7, 10, 102–3, 137, 210, 218, 240, 245–7; indigenous *vs.* non-indigenous education 220–9; staff interviews 219; vocational *vs.* academic education 229–30
Arctic Monitoring and Assessment Program (AMAP) 29
Arctic multicultural educational space 228
Arctic Pedagogical Model of Energy Literacy **163**, 210, 235, 236, 247
Arctic societies 39; adaptation for, energy transition 38–41, *40*; electricity production 35–9; energy production and 35–9
Arctic Sustainable Development 103
Arctic sustainable energy literacy 67–9
Arctic System Model (ASM) 84
Arctic transformation, political decision-making on 33–5
Ascher, M 220
ASM *see* Arctic System Model (ASM)
Association of World Reindeer Herders (WRH) 43
autonomy 48, 91, 92, 94, 156, 189, 191, 206, 250

Baker, S 219
Barnhardt, R 51, 52, 220–1
Barth, M 57
Baxter, J 12
Beck, D 63
Benhabib, S 49
Berkes, F 39
Blaauw, RJ 38
Borealis University (BU) 19, 119, 120, 229, 231, 234, 243; academic *vs.* vocational course 202–3; capabilities developed in students 149–50; cognitive characteristics 235; connections among factors 126; course objectives *127*, *232*; critical thinking 193; curriculum characteristics **138**; curriculum design 204–6; curriculum improvement 158–9; curriculum review 155–6; distinctive factors of course 238; employability *vs.* knowledge application 128; energy literacy 171, 235–6; features of education *134*, **234**; geography and SD 142; importance of local participation 126, 128; interdisciplinary characteristic 237; intergenerational dimension 187–8; knowledge in international geographical areas 128; structure of energy curriculum 142–3; students capabilities 198; students' perception on course objective 195; sustainable development 189
Boulton, J 83, 85
Bowman, C 85
Bridge, G 6, 31, 162
Brundtland Report 5, 7, 53, 54, 98, 113
BU *see* Borealis University (BU)

CAFF *see* Conservation of Arctic Flora and Fauna (CAFF)
Canada: educational initiatives 50; fossil fuels in 34; hydropower resources 37; indigenous peoples 45, 46; provincial governments 34; sustainable development 113
Canadian Arctic 36–7, 39
capability theory 86, 210; capability approach 87–8, 90; core capabilities 89, **89–90**; curriculum planning 91; functionings 88–9; geo-capabilities 90–1; human development 86–7; interdisciplinary participatory pedagogies 92
case study selection 11–15; boundaries and definitions *12*; criteria applied to score **13**, 13–15; data analysis 20–3; education and training 13; impacts of change 13; nature of existent resources 13; qualitative data collection methods 15–20; research & development 13
CAVIAR (Community Adaptation and Vulnerability in Arctic Regions) 41

Index

CBD *see* Convention on Biological Diversity (CBD)
climate change 9, 117, 231; Arctic 5, 36; fossil fuels and 36; and modernization 42
coal 2, 4, 9, 29, 36
Coles, C 137
Colorado's National Snow and Ice Data Center 43
complexity theory 103, 210; arctic environment 82, 83–4; Darwin's ideas and 80–1; definition 80; dynamic nature of change 79–2; holistic education 82–3; knowledge derived from 83; on knowledge systems and curriculum 84–5; mechanical perspective 79; powerful knowledge 85–6
Conservation of Arctic Flora and Fauna (CAFF) 12
Convention on Biological Diversity (CBD) 57, 224
corporate social responsibility (CSR) 54
Cortese, A 232
Cotton, D 67
Cowan, C 63
critical thinking 121–3, 192, 199, 232; Borealis University 193; Glacial University 192–3; Ice University 194; Juniper University 194
crude oil 3, 34
CSR *see* corporate social responsibility (CSR)
curriculum design 153, 239–40; Borealis University 204–6; curriculum improvement 157–61; Glacial University 203–4; Ice University 207–8; Juniper University 206–7; and review 153–6; staff and student meeting 153–6; via module evaluation 156–7
curriculum improvement 157–61; Borealis University 158–9; Glacial University 157–8; Ice University 159–60; Juniper University 160–1
curriculum planning, developmental model of 10, 91
curriculum review: Borealis University 155; Glacial University 154; Ice University 207; Juniper University 206; via module evaluation 156–7

Darwin, Charles 49, 80, 81
data analysis 20–3; comparison 21; document analysis 22; holistic-inductive comparative design 20–1; internal triangulation 22; official documents 21
Delors, J 97, 128, 186, 241, 243, 250
DeWaters, JE 67, 92, **93**, **94**, 235
diesel 32–4, 37, 38, 69
documentary research 20; comparative method 21; holistic-inductive comparative design 20–1; official documents 21
Doll, WA 84
Drake, Edwin 3

EALÁT (Reindeer Herders Vulnerability Network Study) 43, 220, 227
eco-literacy 66, 92, 95, 242, 244
'The Economics of Climate Change' 36
ecosystems 1, 2, 4, 7, 13, 36, 38, 39, 82, 84, 98, 103, 113, 116, 211
education: adaptation tool in changing Arctic 41–4; ESD initiatives in 57–61; in multicultural societies 44–53; for sustainable development 53–67, 55
Education for Arctic Citizenship 174, 178, 181, 210, 241–4, 246, 247
Education for Global Citizenship 128, 242–3
education for sustainable development (ESD) 16, 18, 19, 56, 57, 67, 79, 103, 115, 162, 213, 218, 219, 236, 247; achieving SDGs through 214–16; adherence comparative assessment 182; adherence parameters analytical model **178–81**; adherence to 172–3, **173–4**, 241–5; assess levels of 172; capability theory 86–94; complexity theory 79–86; in educational institutions 247–8; energy literacy and 15; five pillars of 62–4, 64, 243; global citizenship education 94–8; implementing 57; indigenous *vs.* non-indigenous education 228; initiatives 57–8; learning processes 65; policy dimension and background of **58–60**; political activities in 57; positives and negatives **101**, **172**; research 247–9; sustainable development theory 98–103; two sides of **61**, **101**; type 1 (learning about SD) 241, 244; type 2 (learning for SD) 241–2, 244; type 3

Index   259

(learning as SD) 242, 244; type 4 (Education for Arctic Citizenship) 242, 244, 245; type 5 (Education for Global Citizenship) 242–4; types of **62, 102,** 241–5; vision 57
Eglash, R 220
EL *see* energy literacy (EL)
electricity generation 36–7
electricity production: affect Arctic societies 35–9; Norway 33
Elkington, J 54, 55, 99
ELOKA 43, 220, 227
energy 1, 11, 218, 246; Arctic and 29–31; definition 1; demand 4, 5, 9, 38, 230; discoveries 3; education 7, 10, 11, 20, 21, 29, 33, 34, 67, 69, 211, 232; expansion or demand 3–4; as geographical concept 6; history of 1–2; natural sciences and 1; source of 2–4; time component 5; transition and human development 2; transition component 5; uses of 1
energy curriculum 211; common factors 236; distinctive factors 237–9; interdisciplinary characteristic 237; research 237; resources/environmental management 237; structure of 236–7, 239; within sustainability 230–4
energy efficiency programmes 5
energy efficiency technologies 119
energy issues, students awareness of 114–17
energy learning 39, 68
energy literacy (EL) 5, 8, 10, 11, 67–9, 92, 115, 145, 162, 164, 184, 191, 210, 211, 218, 246, 248, 250; Arctic Pedagogical Model of **163**; Borealis University 171; as component of SDL 68; conceptualizing 67; dynamics 68; ESD and 15; geo-capabilities and 92–4; Glacial University 170–1; Ice University 171; improving students' 164; innovative Arctic Pedagogical Model of **165–9,** *170*; Juniper University 171; levels of 162, 235–6; low levels of 68; pedagogical model 92, **93, 94**; pilot projects related to 67
Energy Literacy Framework 67
energy production 2, 5, 13, 28, 32, 66, 79, 82, 118, 197, 237, 249; affect Arctic societies 35–9; environmental impacts of 6; intercultural understanding of 18; societal understanding of 10

energy security 10, 211
energy transition 6, 7, 28, 137, 162, 211, 218, 249; adaptation for Arctic societies 38–41; governance and 31–3; and human development 2
ESD Adherence Comparative Assessment 244
ESD Adherence Comparative Assessment Graph 247
ESD Adherence Parameters 210
Eurobarometer 214
Evans, M 11
Everett, R 4

First Nations 45, **46**, 52, 220, 227, 251; education 220; language 45
First Nations Information Governance Centre (FNIGC) 45
Fish, D 137
FNIGC *see* First Nations Information Governance Centre (FNIGC)
Fondahl, G 44
Forbes, BC 36
fossil fuels 4, 9, 38, 119; in Arctic 33; in Canada 34; climate change and 36
Freire, P 51, 235, 240, 250
Fullan, M 84
functionings 88–9, 228, 247, 250, 251

Galvin, K 16
GAP *see* Global Action Programme (GAP)
GCE *see* global citizenship education (GCE)
GDP *see* Gross Domestic Product (GDP)
geo-capabilities 90; definition 91; in education 91; energy literacy and 92–4, **93, 94**
GHGs *see* greenhouse gases (GHGs)
Glacial University (GU) 19, 111, 116, 119, 221, 225, 226, 234, 243; academic *vs.* vocational course 202–3; capabilities developed in students 147–9; cognitive characteristics 235; conservation of biological diversity 125; course objectives *127, 232*; critical thinking 192–3; curriculum characteristics **138**; curriculum design 153–4, 203–4; curriculum improvement 157–8; curriculum review 154–5; distinctive factors of course 237–8; documentation of traditional knowledge 124–5; energy

curriculum structure 140–1; energy literacy 170–1, 235–6; features of education **134, 234**; indigenous curriculum 141; interdisciplinary characteristic 237; intergenerational dimension 185–6; knowledge application 148; knowledge at 124; knowledge disseminated in communities 125, 126; students at 141; students capabilities 197; students' perception on course objective 194–5; sustainable development 189; traditional knowledge 123, 140–1; untraditional components 140
Global Action Programme (GAP) 57
global citizen learner 97
global citizenship 95; eco-literacy 95; at Juniper University 128; responsibility and ethics 95; self-identification 95; theories of 95–6
global citizenship education (GCE) 94–5, 103, 210; bioregional model 98; domains of learning 97; eco-literacy 95; education for sustainable development 96–8; global citizenship 95; learning objectives **96**, 97
global competence 96
global consciousness 95, 96
Globescan 214
Gough, SR 61, 102, 172–4, 178, 241
greenhouse gases (GHGs) 4, 5, 36, 117, 119
Greenland Ice Sheet 5–6
green revolution 2
Gross Domestic Product (GDP) 4, 5
grounded theory 17–18
GU *see* Glacial University (GU)

Haigh, M 10, 63, 65, 174, 186, 241
Haq, M 87
Healey, M 138
Hein, G 93, 241
Helm, D 9
Henderson, H 241
Higher Education for Sustainable Development 216
Hill, JD 128
holism 82, 103, 252
holistic education, dynamic nature of change 82–3
holistic-inductive comparative design 20–1
Holloway, I 16

Hove, H 56
Hovelsrud, G 39
human development 103; capability theory and 86–7; energy transition and 2
Human Development Report 87
hydrocarbons 4, 51, 158
hydropower 32, 37; in Alaska 37; in Canada 37; Norwegian 33; projects 32

Ice University (IU) 19, 119, 120, 221, 236, 243; academic *vs.* vocational course 202–3; approach at 130; behavioural features 235–6; capabilities developed in students 151–2; cognitive characteristics 235; course objectives *131, 232*; critical thinking 194; curriculum characteristics **138**; curriculum design 156, 207–8; curriculum improvement 159–60; distinctive factors of course 239; energy knowledge and information 130; energy literacy 145, 171, 235–6; features of education **134, 234**; interdisciplinary characteristic 237; intergenerational dimension 188–9; structure of energy curriculum 145–7; student engagement 146; students capabilities 200–2; students design projects 147; students' perception on course objective 197; sustainable development 191–2
Ikeda, D 241
indigenous and non-indigenous students' perspectives 184; academic *vs.* vocational 202–3; capabilities developed from 197–202; critical thinking 192–4; curricula design process 203–8; intergenerational dimension 185–9; objectives of courses as learners 194–7; students' code names **184–5**; sustainable development as practice 189–92
indigenous peoples: Alaska 44, 46; Arctic 35, 46; Canada 45, 46; education system 49; indigenous research 49–50; rights to land 50–1; United Nations Declaration on Indigenous Peoples 48
'Indigenous Peoples in the New Arctic' 44
indigenous *vs.* non-indigenous education 219, 220; biological diversity 227; community engagement 221; ESD 228;

First Nations education 220–1; indigenous knowledge 225; indigenous lecturers 222, 223; intercultural learning 220; multicultural educational spaces 228; natural resource management 223–4; sustainable development 221–2, 224; tradition knowledge 225–7; Western education system 223
industrial economy 2
Industrial Revolution 1, 2, 5, 48, 79
innovative Arctic Pedagogical Model **165–9**, *170*
intercultural education 63
intergenerational component 120–1
intergenerational dimension of SD 185–9, 231; Borealis University (BU) 187–8; Glacial University (GU) 185–6; Ice University 188–9; Juniper University (JU) 188
International Labour Organization's Convention 169 50
interviews: areas covered by 18–19; geographical location of 18; qualitative coding of 18; qualitative semi-structured 20
Inuit Circumpolar Council 213
Inuit lifestyle 39–41
Inuit Local Observations 43
inuits 43
IU *see* Ice University (IU)

James, W 22
Johnson, J 11
Jolly, D 39
JU *see* Juniper University (JU)
Juniper University (JU) 19, 119, 120, 122, 221, 232, 243; academic *vs.* vocational course 202–3; capabilities developed in students 150–1; climate change 143–4; cognitive characteristics 235; course objectives *131*, *232*; critical thinking 194, 199; curriculum characteristics **138**; curriculum design 206–7; curriculum improvement 160–1; distinctive factors of course 238–9; energy curriculum 156–7; energy dimensions 130; energy industry 130; energy literacy 171, 235–6; features of education **134**, **234**; global citizenship at 128; interdisciplinary characteristic 237; intergenerational dimension 188;

structure of energy curriculum 143–5; students capabilities 198–200; students' perception on course objective 196; sustainability programme 144; sustainable development 189–91

Kawagley, OA 220–1
Kelly, AV 91
Kelly, PM 39
Keskitalo, JH 48
Kirkness, VJ 51, 52
knowledge 242; production 243–4; systems 214
Knowledge of the Arctic 43
Kolb, DA 241

Lambert, D 81
'Learning: The Treasure Within' 97
lecturers 19–20
Longman, M 49, 50, 225

Magni, G 223
Marsik, T 37
Martínez-Cobo, J 42
Mason, M 80
McCann, H 231
MDGs *see* Millennium Development Goals (MDGs)
Meehan, RH 36
memes 63
Mezirow, J 63, 93, 241
Millennium Development Goals (MDGs) 54, 211, 214
Morrison, K 80
Moses, J 82
multicultural education 52, 228, 245, 251, 252

National Energy Technology Laboratory (NETL) 32
natural gas 3, 9, 29, 32–4, 36, 37
natural resources management 111, 171, 223–4, 242; students awareness of 118–20, 231
NETL *see* National Energy Technology Laboratory (NETL)
Northwest Territories: diesel 32–3; First Nations **46**
Norway: electricity market 36–7; electricity production 33; energy system 33–4; natural resources 41
nuclear energy 4

Nunavut 32, 44, 45
Nussbaum, M 66, 89, 91, 99, 186, 250

OCS *see* Outer Continental Shelf (OCS)
oil: in Arctic Alaska 31; geologic provinces 29
'The Origin of Species' (Darwin) 80
Orr, DW 223
'Our Common Future' 4, 53
Outer Continental Shelf (OCS) 9, 29

Parkhe, A 17
petroleum formation 2–3
Petrov, A 66, 100
Poppel, B 66, 223
powerful knowledge 85–6, 103
Powers, SE 67, 92, 235
Prigogine, Ilya 80
Prudhoe Bay 31

qualitative data collection methods 15; areas covered by interviews 18–19; interviews 18; lecturers 19–20; methodological procedure 16–18; students 20
qualitative research 11, 16, 17, 20, 49, 108, 171
quantitative research 16

RCE *see* Regional Centres of Expertise on Education for Sustainable Development (RCE)
RE *see* renewable energy (RE)
Regional Centres of Expertise on Education for Sustainable Development (RCE) 53
renewable energy (RE) 5, 7, 13, 32–4, 37, 38, 67, 69, 118, 119, 130, 147, 151, 171, 191, 236
'Renewable Energy for the Arctic: New Perspectives' 69
research, structural course component 138–9
residential schools 70n1
resources/environmental management 139–40
resources management 237
Rio+20 Conference 57
Rio Declaration on Environment and Development 53
Roberts, A 84

rock oil 2
Russian Federation 44
Ryan, A 56, 65, 92, 218

Sámis 43, 46, 48; indigenous education system 49; in Norway 45
SATVA *see* Student and Teacher Value Assimilation (SATVA)
Scandinavian Arctic 36
Scott, WAH 61, 100, 102, 172–4, 178, 241
SDGs *see* Sustainable Development Goals (SDGs)
SDL *see* Sustainable Development Literacy (SDL)
SEL *see* sustainable energy literacy (SEL)
Sen, A 99, 250
Sen, Amartya 87–9
SES *see* sustainable energy system (SES)
Shell Oil Company 54
Shell Report 54
Sillitoe, P 223
SLiCA (Survey of Living Conditions in the Arctic) 41
Smit, B 38, 39
Smuts, J 82
social constructivism 108
social rights 48
social science investigation 18–19
societal structures 11
St. Clair, R 93, 241
Sterling, S 56, 61
Stern, NH 9, 36
Student and Teacher Value Assimilation (SATVA) 53
Subarctic regions: boundaries and definitions for 12
subsistence farming 2
sustainability 55–6; cultural value 116; definition 99; spatio-temporal dimensions of 8; as system 99–100
sustainability education 61
Sustainable Cleveland 55
sustainable development 111, 115, 208; biological diversity 125; Borealis University 189; economic component of 192; energy issues 115–16; Glacial University 189; Ice University 191–2; intergenerational component 113; Juniper University 189–91; local actionable scheme 117; as political concept 112; students awareness of

114–17; understanding concept of 111–14
sustainable development (SD) 1, 7, 53, 143, 184, 218, 219, 234, 240; biological diversity 227, 230; critical thinking 232; educational opportunity 219; education for 10, 53–67, 55; and energy 226–7; environmental aspect 54; geography and 142; interactions among natural resources 231; intergenerational dimension *see* intergenerational dimension of SD); intergenerational impacts 231; literacy 66; local reality of 65; natural resources management 231; pedagogical systems of teaching 173; as practice 189–92; principles of 18; raising awareness pattern 230–1
Sustainable Development Goals (SDGs) 54, 57, 210, 212, 214; achieving, through ESD 214–16; commitment to 213; global goals 211–12; siloed approach to development 213; traditional knowledge and 213
sustainable development literacy (SDL) 4, 8, 66, 67, 92, 100
sustainable development theory: cultural factor and 98–103; education and 98–103
sustainable energy literacy (SEL) 69
sustainable energy model 10, 67, 69, 134, 210, 218, 235, 236
sustainable energy system (SES) 4, 5, 23, 31, 153, 218, 230; educational initiatives 69; multicultural perspectives of 245–7
'Sylvicultura Oeconomica' 99
systemic thinking 16, 144, 193, 228, 238
Systems Change Model 53

Taylor, C. 19
TEK *see* traditional environmental knowledge (TEK)
Thomas, G 17
'Three P formulation' 54
Tilbury, D 10, 57
traditional environmental knowledge (TEK) 39
tradition knowledge (TK) 42, 43, 123, 213, 224, 225, 250, 251; documentation of 124–5, 210, 226, 227

transformative learning 61, 63, 65, 235
'The Treasure Within' 243
triangulation 22
Triple Bottom Line (TBL) 28, 54, 65, 99, 245, 249, 251; agenda 55; drivers of 55; sustainability 55–6
Tuhiwai Smith, L 228
Tuktoyaktuk 40, 39

Ulukhaktok 40, 39
UN Conference on Environment and Development (UNCED) 98
UNESCO 56, 243
UNESCO ESD 21, 53
UNESCO Global Citizen Education Framework 96
UNFCCC *see* UN Framework Convention on Climate Change (UNFCCC)
UN Framework Convention on Climate Change (UNFCCC) 38, 57
UN General Assembly 48, 67
United Nations 211, 212, 215; Brundtland Commission 4, 5; World Decade on Education for Sustainable Development (2005–2014) 57
United Nations Conference for Environment and Development 53, 54
United Nations Declaration on Indigenous Peoples 48
United Nations Educational, Scientific and Cultural Organization (UNESCO) 57
United Nations General Assembly 53
UN SDGs, relevance to Arctic 210–14
UNWCED *see* UN World Commission on Environment and Development (UNWCED)
UN World Commission on Environment and Development (UNWCED) 53
U.S. Geological Survey 9
US National Education Standards 67

van der Horst, D 92, 137, 164
Vare, P 100, 172, 241
vocational *vs.* academic education 218, 229–30
vulnerability 38, 39–41, 50, 215, 250

Walkington, H 174, 235, 241
Wandel, J 38
WCED (The World Commission on Environment and Development) 98–99
Western educational system 51

Wiltse, N 37
World Bank 54

Young, M 52–3, 81, 85
Yukon First Nations **46**

Printed in the United States
By Bookmasters